Treehouses
in Paradise

Treehouses in Paradise

Fantasy Designs for the 21st Century

David Greenberg

Abrams, New York

Top Ten Finalists

Contents

Introduction
David Greenberg

I. THE COMPETITION

I cannot remember the exact moment the idea for a treehouse competition came to me, but I know why I thought of it. I had unwittingly landed far more contracts to design and build treehouses than I knew what to do with. It was 2000 and I was expected to deliver designs for more than one-hundred treehouses located in various sites around the globe, including a virgin beach in Fiji, a lush rain forest in Hawaii, a sand dune in the South China Sea, and a spectacular lagoon in Vietnam. I couldn't imagine designing so many by myself and certainly didn't have the money to hire so many architects. By the time the competition kicked off, four-hundred architects had signed up.

The official host site (www.arcspace.com) was set up in Denmark, thanks to Kirsten Kiser, who runs this weekly architectural magazine and Internet archives. Soon after the launch in September 2000, the general concept and rules were picked up by a variety of architecture-related sites, including www.deathbyarchitecture.com, which specializes in architectural competitions, and www.designarchitecture.com. Even print media picked it up, including the *New York Times*. Registration emails from all over the world started pouring in. By deadline time (January 15, 2001), architects from 38 countries had entered. The prizes being offered to the winners were quite small—some tiny cash prizes and the offer of a free month's stay in one of my treehouses in Hawaii. The real reward was in the challenge itself—to create an innovative structure under an unusual set of restrictions, set in a unique environment. It was also freeing—few if any walls were required, no heat or cooling mechanism would be necessary, and there were no building codes to adhere to. Many participated for no other reason than for the sheer fun of it.

One of my favorite jobs was opening the fifteen or so packages that streamed

OPPOSITE, CLOCKWISE FROM TOP LEFT: The author's treehouse in Hana, Maui, Hawaii; jury member Master Chang overlooking Nanshan Buddhist Culture Zone in Hainan, China; a beach at Nanshan; bay at Quemea Island, Fiji; lagoon at Lang Co, Hue, Vietnam. All photos by the author.

in daily to our makeshift office for weeks on end. They arrived like hundreds of Christmas presents from around the world. I carefully excavated each package—typically triangular-shaped rolled drawings or large, flat packages, often 25 x 40 inches—and then excitedly ran around showing them to friends at Perloff/Webster, whose office space in Venice, California, I was borrowing. I especially began to look forward to the entries from Germany. There were so many of them! A couple of German universities entered whole classes. Some of the student models were outrageous and perfect at the same time. Looking back, we should have allowed ten-year-olds to compete—they are the ones still climbing trees, which is what it is all about, after all!

One of the greatest payoffs for me was listening to the top architects, artists, and designers who made up the jury talk about my favorite subject: treehouses. The jurors—consisting of a few architects, some designers, a landscape architect,

and a handful of artists—was led by Elyse Grinstein, an architect and patron of the arts in Los Angeles who is famous for her artists' salon in Brentwood. Another member of the jury was Ming Fung, director of the graduate program of architecture at the Southern California Institute of Achitecture (SCI-Arc), who is also on the board for the National Endowment for the Arts. She is now a partner at Hodgetts and Fung in Los Angeles. Additional jury members were sculptor Ken Price and artist Billy Al Bengston, who had particularly interesting perspectives. As could be expected, they seemed to enjoy the treehouses that most strongly resembled works of art. One of their favorite entries, for example, was reminiscent of toy airplanes and other structures that could have been designed by a five-year-old (page 144). (That particular architect apologized for the presentation—apparently all the detail drawings were destroyed in an unfortunate boating accident.)

After reviewing the designs, the jury selected one-hundred "winners," as well as ten finalists. After some debate, it was decided that the one-hundred winners would decide who would be awarded first, second, and third place, to be chosen from the ten finalists. The top ten finalists' works were shown on my Web site (www.treehousesofhawaii.com) and voting took about a month. It nearly turned into a five-way tie for second place.

Later in 2001, I visited several of the finalists, some of whom were in Germany, including a group called Drechsel Fromme Architekten, who did a fantastically original design (page 82) that strongly resembles a hideout for an extra in *Road Warrior*—a sort of collaged spiraling thing held up by giant bamboo. I had them accompany me on some newspaper interviews about the competition. In East Berlin I sought out Marcus Bach, who had designed a treehouse made entirely of bamboo (page 30). This year I had him help me out with

a master plan for an ecotourist resort I've been hired to design in China that is contained within a Buddhist theme park.

One of the best designs, in my opinion, came from Kendel Architekten in Berlin (page 106). The red-and-silver high-tech hammock looked something like Santa's sleigh. Winning didn't matter much to Jan Kendel, who had a great attitude about the contest. As far as he was concerned, at the very least his design could be used on the company Christmas card.

After visiting more architects in Frankfurt, Stuttgart, and Munich I headed to Zurich to meet with Mireille Turin and her partner, Peter Sigrist, who won first place with their colorful tentlike structures suspended over a desert island (page 16). I particularly liked this one as it seemed like the perfect solution for the globe-trekking nomads that I was hoping to accommodate with my treehouses around the world. In this case, the structure was designed to be collapsible and transportable—you could fold it up, put it on a plane as baggage, and take it anywhere. Ms. Turin and I talked a great deal about mass-producing her design using state-of-the-art windsurfing technology, most of which actually comes from Japan, Taiwan, and China even though Maui is considered the leader in the field.

By far one of the greatest rewards of the competition, at least for me, was when the winners came to stay in my treehouses. Marc Ackerson and Oisin Clancy of the group called Fieldlines, with offices in New York City and work in Ireland, showed up with about ten friends, including an eccentric astrophysicist. Their treehouse took second place, and they helped me out with some designs for a variety of projects I was working on in China that year.

If the winners still hadn't fully wrapped their heads around the essence of tree-house design through the competition, they probably had their "a-ha" moment when they arrived at my tree-houses in Hawaii. Designing for "man in paradise"

involves a different process than designing a house or a building located in a city, which is either trying to compete with the buildings around it or trying to fit in. In paradise, the only competition is Mother Nature, and who wants to compete with perfection? Design in nature is a formidable task.

This is a book about creativity, and creativity requires a spark of some kind. In this case I think it was both the fantasy of a treehouse and the fantasy of a tropical paradise that dazzled close to five-hundred architects from 38 countries. Architects experience as much trouble thinking "out of the box" as anyone. Put an architect in a place of nature where boxes don't work and voila!—A chance for creativity is ignited.

Paradise is defined in the dictionary as a place, situation, or condition of perfect happiness, as well as a place where there is everything that a particular person needs for his or her interests. The program for

this competition was simple: to design the
perfect structure in the perfect place where
everything you need is accessible. This book
is a testament to the vivid imaginations
of all those architects who took up the
challenge and revealed a remarkable ability
to form images and ideas of things never
directly experienced in day-to-day life.

Competition Guidelines

This is the official announcement for the international competition sponsored by Treehouses of Hawaii, Inc., and www.arspace. com. It was launched in September of 2000 with a deadline of January 15, 2001.

DESIGN A TREEHOUSE FOR A TROPICAL ISLAND RESORT

The competition sites are near the beaches of Fiji, Vietnam, China, and Maui, where there are no existing trees. The majority of the treehouses will be in coconut palm trees that will be brought to the site and planted as per the design. The tree size is the designer's choice; they can vary from ten feet to fourty-five feet in height.

You must consider that all palm trees grow from the top up; a construction in a palm tree two feet from the ground will never move up.

All connections to a tree must be in the form of a stainless-steel sleeve about $3/16$-inch thick and about ten inches high.

For the high decks, it is not a good idea to make rigid connections between all the trees as the trees can be blown in different directions and fight each other in high winds.

The sites in Southeast Asia are liable to have typhoons and heavy monsoon rains, which means you must design to resist heavy wind loads. Basically, you must use natural materials but, although tin is not natural, corrugated tin is natural in the jungle as roofing material because of the rain. You can also choose to design a roof of vinyl/plastic that is stapled on and replaced every year. This kind of roof also serves to let in light in the dark areas under the tree canopies.

The sites and climate allow for a close relationship to nature and an openness not normally allowed in construction. There is almost no need for walls; only for screening materials which are available locally in beautiful colors.

OPPOSITE, CLOCKWISE FROM TOP LEFT: "Treehouses of Hawaii," the author's company, at the entrance to Nanshan Buddhist Culture Zone in Hainan, China; lagoon at Lang Co, Hue, Vietnam; exhibition of projects from the treehouse competition ready for the jury's review at Frank Gehry's Edgemar Center in Santa Monica, California; first-place winners Mireille Turin and Peter Sigrist in Zurich, Switzerland; the author with three of the Swedish architects who designed treehouse 85, pages 134–35. All photos by the author.

The total floor area should be between five-hundred and one-thousand square feet and should accommodate two to six people. No structural elements may touch the ground except for access stairways.

For the construction there are several varieties of hardwoods in China, Vietnam, and Fiji. In Hawaii you have to be more creative.

The designs must include a toilet and a shower of ecological nature. The shower will have propane water heaters, the toilets will for the most part be compostable, as most sites may not have either sewage or septic systems. You can incorporate the ground-floor area for cooking and showers.

Most sites will not have electricity, even for construction. However, since good handicraftsmen are plentiful, designs that emphasize craftsmanship are encouraged.

Building codes for treehouses have not yet been regulated in any of these countries. The winning treehouses will be adjusted, when necessary, to make them safe for human habitation.

The entries must include plans, sections, and elevations as well as sketches, significant details, and any written information you may deem important.

The winners will be asked to submit a scale model.

The winning designs will be judged by firmness, commodity, and delight.

You can find plenty of resources if you search "treehouses" on your favorite search engine. You can start off with "Treehouses International" and "Treehouse Out 'n' About."

The designers of those treehouses selected for the top one hundred will win a free month's stay at one of the author's treehouses in Hawaii or China. Winners of first, second, and third place will win, in addition to a free stay, free roundtrip airfare.

The Jury

Elyse Grinstein
architect and patron of the arts

Ming Fung
architect/partner, Hodgetts and Fung
director of the graduate program, SCI-Arc
Board of Directors, National Endowment
for the Arts

Deborah Sussman
designer

Pamela Burton
landscape architect

Stephen F. Jones
architect

Billy Al Bengston
artist

Ken Price
sculptor

James Ganzer
artist

Michael Garnier
president, World Treehouse Organization

David Greenberg
urban designer

Master Chang
Zen Buddhist monk

Kirsten Kiser
director, www.arcspace.com

A NOTE ABOUT THE ENTRIES:
The order of appearance of the designs in this book has nothing to do with their excellence or placement in the competition. The designs are sequenced to emphasize the dazzling array of concepts, the visual interest of the executed designs, and the variety of countries from which the treehouse plans were submitted.

Mireille Turin and Peter Sigrist

01 Switzerland
Top Ten Finalist, First Place

The news of the treehouse design competition sent me back to childhood memories—climbing trees in my village, feeling aloft and detached from the world in a very special way. The sensation is similar when entering a church, or one of Louis Kahn's buildings in Bangladesh. Sitting on the tree, one can focus. I loved to sit in all sorts of trees, some very high, some embracing. Each tree has its ramifications, so the issue was to think about building with palm trees. I did not want to build an actual house, but rather a platform that brings back the sensation of my childhood experiences. I arranged the platform on the trunk using the structure of an inverse umbrella. Peter came up with the conical shape to cover the bottom platform that could hang from anything that was above ground. It was designed with an inflatable base of tarpaulin, with tarp-batons welded to the inside for support. The textile hull is conical.

The treehouse opens and shuts like an umbrella, so that a traveler can take it away with them and set it up at the next location. The stainless-steel sleeve is loosely attached to the palm tree on the ground. The steel beams are connected and leveled out from the palm with 20mm screws on top of the tree, and another steel sleeve with a pulley is ready to haul the whole construction approximately ten feet above the ground, at which time the stainless-steel sleeve is properly fixed. The steel skeleton is covered with colored canvas; the canvas also works as flooring material, similar to a catamaran's deck. Curtains subdivide the circular floor plan, providing polyvalent flexible space for living and sleeping. Facilities for cooking and hygiene are arranged around the tree trunk. A rope ladder gives access to the main floor. The tentlike double-layered canvas can be pulled up to allow light inside. Futons serve as furniture, either spread, rolled up, or stowed away.

PRIVACY—TEMPORALITY—NOMADS

"I awoke this morning at 6:00 A.M. with a slight feeling of eternity. As I walked toward the shore, I still remembered the million beams of light surrounding my private castle in the sky."

This proposal can be imagined as a camp in paradise, stimulating feelings of temporality, adventure, interaction, a life detached. From afar, the camp looks like a shiny field of colored plants. The

treehouses appear as a field in bloom, while the human beings within them celebrate passion and reflection.

COLLECTIVE DREAMING

"I used to dream of a place detached, high above the ground, a place of my purest childhood images of independence, truth, and desire."

By means of the translucent skins, the conditions for both private and collective dreaming can be generated.

SKIN AND STRUCTURE

"I rediscovered the most primitive form of seduction as I looked through a landscape inhabited by shiny bodies, a field of moving silhouettes."

THE NEXT GENERATION

"I was struggling against my primary needs in a desert somewhere near Morocco, lost in an endless topography, when I blew up this strange thing made of air and plastic, giving me shelter. A complete home ready in three seconds."

This is the ultimate tool for the lonely nomad, traveling the earth's surface with his home on his back, folded into a three-quarter-pound package.

What happens when the ground—geographical, geological, cultural, economical—becomes distorted through mechanisms of temporal and spatial displacement that characterize our age? How can we frame architecture within these increasingly shifting conditions of the ground? We can no longer rely on the classical relationships between building and ground.

NOTE: *This design will be available as a prefabricated treehouse using windsurf technology and materials. For more information contact the author (see page 156).*

JURY REMARKS

BURTON: The hanging lanterns of Babylon, or Hawaii. I think the idea of the scale is what really works. I love it.

FUNG: All the jury likes this very creative way. It is inventive, using such simple technology as an umbrella; borrowing from another system. It is so whimsical with such a light design touch. It makes you want to be in there.

Barbara Boegershausen

02 Germany

This treehouse tube is designed for four coconut trees 33 feet high; it can accommodate four persons.

This lightweight timber tube made of stacked planks is a stable construction that resists heavy winds and can span a distance between the pairs of palm trees of approximately 37 feet without using big girders. It is made of crossed ribs built up of stacked and screwed planks (4 x 2 cm/8 cm) and an additional layer of stiffening boards on the exterior. The shell is made of simple beams that can be built up layer by layer and screwed together. The stiffening planks on top and bottom of the tube allow a view outside without losing the shade provided by the boards. Fly screens may be installed on the ribs before the planks are installed. The end façades are like bookshelves with sliding doors to the terrace. The fillings can be made of locally available screening material or single glazing.

Martin Spuhler and Markus Horn

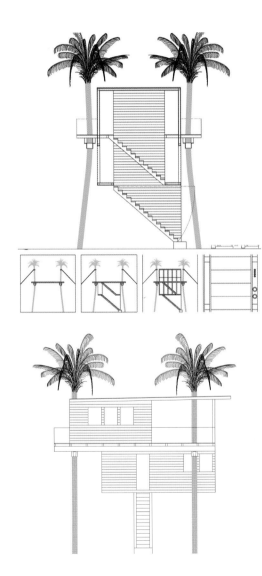

As the treehouse developed, we felt transported into our childhood. We remembered the freedom and adventure as well as the anxiety and the need for protection.

The defining factor of the design was the stipulation that only the stairs could touch the ground. Considering that the tree trunk would grow in circumference, we developed a flexible cuff for the mounting of the carrier console.

The four palm trees are set in a slanting position before the mounting of the house, in order to prevent subsiding of the structure. Prefabricated wooden plates form the interior. The slightly slanted roof can be covered with sheet metal or foil. The living area is distributed on two planes: bedrooms with adjacent bathrooms on the lower floor, and living rooms and kitchen on the upper floor. The staircase can be pulled up for security. We see the structure as a colored object set in a tree.

Alvaro J. Fernandez-Velasco
04 Argentina

HIBISCUS TREEHOUSE

The hibiscus, state flower of Hawaii, seems to be the perfect answer for this treehouse project. The design consists of "petals" that grow on each level; some of them are open, some are closed, and some are no longer there.

Kitchen and bathroom are on the ground level, avoiding the problem of pipes on the palm trunk carrying plumbing to higher levels.

The rest level is located on the first floor; it's the most protected area of the treehouse. Between the ceiling and the bamboo flooring—above as well as inside the double layer bamboo walls—there is a concealed polyethylene waterproof barrier. The walls are oriented to the prevailing winds, which will be deflected by their curved shape.

The living level has more openings to the outside with an excellent view, and is furnished with local chairs and tables.

At the top, the finishing touch is the reading terrace, which has the best view and full contact with palm fronds and coconuts. Lighting will be supplied by oil lamps.

Kai Wei Bin

The shape of a boat can be a space in which to float away from the daily struggles to a place where nothing matters but the sea, the sky, and me. A tree is solid, stable, and rooted. This design gives an added dimension—a feeling of weightlessness as if floating on a warm, fluffy cloud.

By crisscrossing the palm trees to cradle the treehouse, we have arrived at an elegant solution for elevating the space.

All islanders possess a rich tradition in boat-building techniques; this design can be seen as an extension of those traditions. The smooth surface of the planking, normally attached to the outside of the boat's ribs to enable it to slice efficiently through the water is reversed, because we need the smooth surface in the interior. This simple "turning a boat inside out" gesture makes the difference between a boat meant for the water and a boat in the sky.

Can the culture of the natives grow and mature instead of adopting the invading culture? May we all enjoy an island while being always mindful that we are altering the conditions that have existed for centuries? May we question ourselves about every action we take, regarding the repercussions?

Augustin M. Ceballos Ruiz
06 Mexico

This unique structure is more like a tent in a hammock than most treehouses we know. It is simply rendered in black-and-white and quite genius, actually. The perfect exit could be a canvas ramp slide which would only leave access as a major problem.

GREENBERG: I am amazed by the beauty of the design with its low-tech rendering by freehand pencil and paper expressing a heavenly hangout.

Robert Walsh

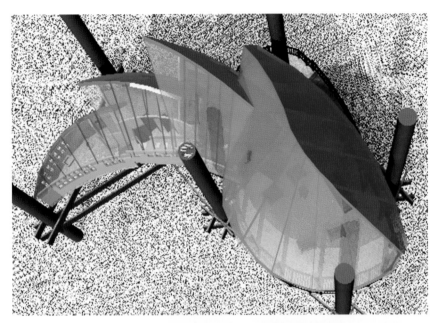

I explored the solutions developed by indigenous cultures that had developed effective typologies for treehouse; I imagined what would it be like to enjoy a vacation in one of these structures, and I returned to the question of the being's character, the sort of animal that this treehouse wanted to be.

During these explorations I observed that in indigenous tropical cultures the reality of the manmade living environment was largely defined by the roof and the floor; walls play a far less crucial role in these traditional tropical dwellings than in our western culture. Sometimes the structure amounted to a roof held aloft by columns. Furthermore, these roofs tended to be grand and majestic structures, the strongest identifying expression of these places. In my project you can see that much of the fun really comes from the curving vaulted form of the roof, and that dispensing with walls in the main living space does not eliminate a feeling of enclosure, although it does permit excellent views.

During this process I became fascinated with bamboo toy snakes of all things. This somewhat surprised me, but I found myself intrigued by the idea of curving structure winding its way through a grove of palm trees, sheltered by the fronds above yet enjoying a commanding view of the landscape beyond. Oddly enough these toy snakes, which are made of sections of bamboo joined together, began to suggest an organizational scheme in which each segment was a separate space linked to the succeeding one. These segments overlapped, changed in height, and also presented the sort of flexible geometry that would allow a wide range of unique configurations depending on the particular circumstances.

The roofs are to be canvas over a wood-and-bamboo frame. The filtered light through a canvas roof seems pleasant. In addition, the canvas could be of different colors to allow a number of treehouses to each have a unique presence. Along a stretch of beach, I envision them being illuminated from within, gently glowing at night like so many paper lanterns.

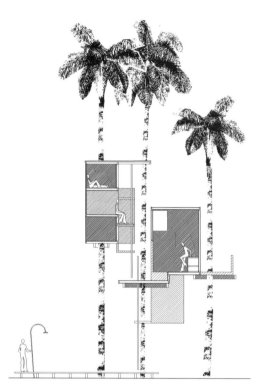

A house in a tree doesn't need spatial relationships on the horizontal plane as houses on the ground generally do. For a more playful spatial configuration, we used a series of palm trees to hold several elements in the air, each intended for a different activity. A void space containing two ladders links them all, becoming the heart of the house by creating circulation.

There are three spaces large enough for two people—to sleep, to keep belongings, and to enjoy the views, but not large enough for walking around. The long side is protected by a mosquito net and a roll-down blind.

There is a living space that includes a kitchen and opens onto a terrace. Underneath there is a well-ventilated chemical toilet.

A wooden deck on the sandy beach signals the entrance to the house and offers a shower before climbing home.

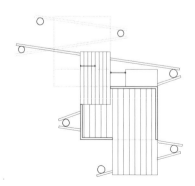

Scott Evans
09 USA

To resort to a treehouse should be a playful thing. Derived from nature (beehive) and an act of man (weaving), this treehouse interlaces itself into the natural environment as an appropriate element to form a shelter within its surroundings. The hive form has been used for centuries to provide shelter to living organisms. Weaving has been used throughout the centuries for utility as well as decoration.

This treehouse rests within a moat sheltering itself in isolation, with only connections to the earth through a vertical pulley lift and slide. The plumbing would be hidden in a mechanical "tree" running from the kitchen/bathroom wall to the ground. This treehouse is approximately 600 square feet and can be opened to 360-degree views or completely closed for privacy. The toilet room and shower become isolation points, as they are the only areas that can be closed off to the inside, yet remain open to the outside.

It is a meditation on the struggle between man and nature.

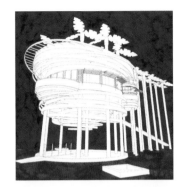

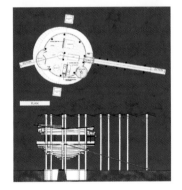

The structural requirements for load-bearing walls have been eliminated because the roof loads are not being supported by the floor. Steel is used for its excellent tensile strength, with wood members for small spans and minimal loads. The simple tensile structural system suspends rather than supports from beneath the floor platform and the roof system. Twelve small suspension cables made of coated steel replace a typical system of walls, posts, floor and roof beams, and floor support brackets. The key structural components are prefabricated, employing high-tech materials to minimize their weight and size and ensure their precision and simplicity of assembly.

The top hanger system (lightweight metal collar, trunk anchor bolts, and cable eyelets with pulleys) is installed on the coconut tree trunk. Suspension cables are threaded through the eyelets secured at ground level. The roof system made of lightweight horizontal members acting in compression is assembled at grade in correct position around the tree trunk. Cable eyelets and hangers are fastened and the suspension cables threaded through them. The assembled roof system is lifted ten feet on suspension cables and temporarily secured in place.

The floor system is assembled on the ground directly below the suspended roof system. Cable hangers are anchored to the key floor joists and suspension cables threaded through and secured in position fixing the floor-to-ceiling height. Pulling the suspension cables lifts up the entire lightweight structure. When the required height is reached, the cables are locked in position at the floor system anchors around the trunk, securing the skeleton in place. All walls are structurally suspended from the roof system and fastened to the floor merely to secure them in place. The flexible plastic plumbing piping is fitted from beneath, and the sun deck, shower enclosure, kitchenette, and propane tank storage pen are built at grade.

The structure sways gently, and there is no acoustical separation from the environment. Large openings provide views in all directions, but the exterior walls can be entirely omitted, or replaced with protective railings. The building's high-tech, prefabricated components contrast with the natural, indigenous materials, enhancing the aesthetic appeal of both.

Cables may be used for cross-bracing and tying the structure to the lower trunk in windy locations.

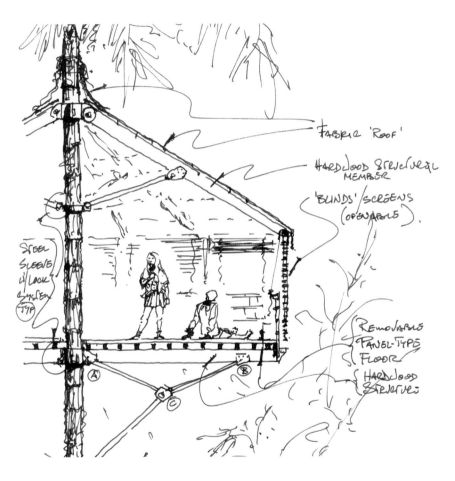

FABRIC 'ROOF'

HARDWOOD STRUCTURAL
MEMBER

'BLINDS'/SCREENS
(OPENABLE).

STEEL
SLEEVE
W/LOCK
SYSTEM
TYP

REMOVABLE
PANEL-TYPE
FLOOR

HARDWOOD
STRUCTURE

Marcus Bach

11 Germany
Top Ten Finalist

THE STRENGTH OF BAMBOO, OR THE SKY AS A TERRACE

The Chinese wise man Pou Sou Tung says: "A meal should contain meat, a house should contain bamboo. Without meat we waste away; without bamboo we lose our cheerful composure."

One of the main reasons for utilizing bamboo as a building material is that bamboo provides unbeatable value for the money. It can replace many tropical woods and thus help protect the rain forest. In addition, bamboo is very cost-effective and easy to work with. It is easily bent, and once bent, serves a wide range of purposes.

The symmetrical form and space of the shown design are the result of the bearing mechanism. An element can bear more load when it's curved, and this principle can also be utilized with bamboo. This brings about another aesthetic element in bamboo construction.

Four coconut palm trees provide support for a form-active structure system of four arches. Only one point is connected to a tree. The other three points are flexible in horizontal directions. Form-active structure systems develop horizontal stresses at their ends. Four crossed bamboo canes receive these stresses, because

bamboo has great tensile strength (twice as much as steel). This structure system is the mechanism for bridging the span of 40 feet.

For connecting the bamboo cane, ropes and cords of bamboo rind or coconut palm fibers are employed. Synthetic materials are also possible. Bamboo cane split into strips forms the building element for screens and wall structures. For reasons of comfort, kitchen, shower, and toilets are to be fixed on the treehouse. Maximum load is concentrated on the space between the palm trees.

The design works with four supporting points, but smaller or even bigger designs are possible, all with the same symmetrical bearing system for the decks. Only the number of overhanging deck parts is flexible. Several different sized treehouses form a little village on a beach, and each of them is equipped with double beds, casual bamboo tables, seating, desks, and high ceilings as the most effective way to cool interiors. The design is based on the idea of prefabrication. There is a primary structure that can be shipped and assembled in single parts, all with the curved form and dimensions, and then there are all the other parts such as furnishings, roof, the shutters, and windows, all prefabricated. The roof is separated

into two levels with space for ventilation in between. Here the symmetrical design corresponds to the idea of prefabrication because most parts have identical dimensions. In case of heavy winds the primary structure system will remain. All other parts can easily be replaced.

For French philosopher Gaston Bachelard, all rooms had lingering dream values. He considered the dreaming man a function of the building. He believed that many dreamers imagined the place. The Belgian painter, René Magritte, sketched the house and the face of a dreamer as a creature with one body and one soul. In the pictures by Magritte, plants penetrate living rooms, a forest is situated on a floor, or the trunk of a huge tree can be opened like a cupboard.

I started with a journey into the simple rooms of my childhood. Rooms that offer shelter look alike in every childhood. They are attics, cellars—the uninhabited places of a house—or the caves in a nearby forest, or the garden treehouses. As one fascinated by a treehouse, I climb the stairs of trees in order to return to the attics and cellars of times long passed, so my house is part of the sky and the whole sky is its terrace.

BURTON: It's taking traditional forms of architecture and applying them to a nontraditional situation and letting the expressions of the materials pull it together. The image is one of the most compelling of all that we have looked at. He has a wonderful sense of composition.

FUNG: It is conceptually very strong, and the architectural forms are generated by the details of the structure.

The whole thing is put together around the central key pole or tree and the form springs out of this.

GREENBERG: This is one of my favorites. I am in the process of having this prefabricated entirely in bamboo in a special factory in Vietnam. Hard to believe that it came out of a studio in Berlin by somebody who had never been to a tropical country.

NOTE: *This design will be available as a pre-fabricated bamboo treehouse in the near future. For more information contact the author (see page 156).*

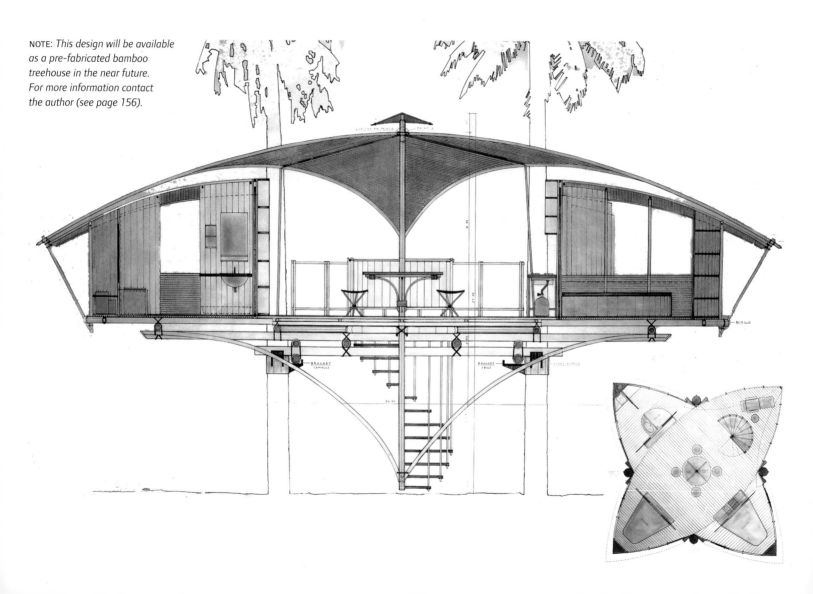

Khanh Phan
12 USA

When I was young, my dad built a treehouse for my siblings and me. Perched in our treehouse retreat, I learned many things about nature and the processes of life. The treehouse, a thing to climb, a space for reflection, and a powerful mechanism for learning an appreciation for nature. My childhood treehouse had a great influence on my decision to become an architect, and the lessons still shape my ethics and beliefs as a professional.

My father constructed the treehouse from materials bought from a local lumberyard and used our apple tree as the support for the platform.

He fastened the eight-by-eight-foot platform to the trunk of the tree using large metal spikes and drove nails through the fragile branches. The natural curves of the tree did not harmonize with the hewn, modular lumber, so my father cut the tree to fit the modular components. Unfortunately, he unknowingly left the cut limbs of the tree exposed, causing the tree's moisture to bleed and causing the tree slowly to die. For years, the apple tree cradled the platform between its branches until, finally, its trunks and branches dried and it could no longer stand.

My treehouse design draws from all of the great aspects of my childhood treehouse and integrates lessons from its demise. The treehouse is a composition of living and fallen trees. I began with the configuration of the old apple tree from my childhood home. The two platforms in the treehouse are triangulated to better fit between the trunks of the living trees.

Fallen trees are added as structural elements, and to create rhythm in the composition. A natural, curving element begins on the ground and forms the handrail, rises through the tree to become the roof structure, then culminates high above the treehouse branches. The multilevel platform configuration creates a space for gathering and a separate space for reflection. The resulting organic form rising from the ground invites one to climb to the heights of its platforms. Unlike my backyard treehouse, the climate allows this design to have only light screens and no walls.

When was the last time you were in a hammock? Why were you there? Where was it? A hammock is a magic tool for memory and dreams, and it's also quite portable. You can change the place and keep the dreams.

The design for this treehouse is a hanging structural surface, like a composite hammock, constructed for a multi-dwelling community.

A lot of dreams are supported between palm trees, between the sound of winds and the shadows of passion. It must be a good environment, with elements such as "pure, clean water" or "far, shallow lagoon." If you get there, a strong golden wind will blow down on your body on the hidden beach in the night.

Don't forget that paradise is linked to the practices of urban living.

Robert Ley and Joshua G. Stein
14 USA

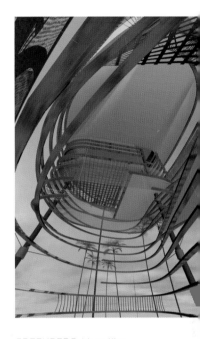

HULA HOUSE

The Hula House attempts to redefine the understanding of "treehouse." The idea of vacation becomes a complex exploration rather than a simple escape.

The Hula House fosters a mode of living that is retuned to the rhythm of the surrounding nature and culture. It avoids the most obvious strategies of treehouse construction—the platform and the pod. The platform maintains that daily life can be duplicated twenty feet off the ground. The pod is suspended from the trees, allowing movement with them, but the relationship to life in the trees remains abstract because of the tendency of the pod to encapsulate space, thereby severing the possibility of truly moving and living among the trees.

The Hula House escapes these pitfalls and simply encourages meaningful life among the trees. It inhabits a sandy stretch of newly planted coconut palms along the beach; it sways with the trees instead of sitting on them. Sometimes surrounding the trees, sometimes loosely suspended among them, each Hula House responds in a slightly different way to its context.

GREENBERG: More like sculpture than architecture, this simple but beautiful design has been chosen by a client in Hainan, China, to be built on a beach.

BURTON: They are using all the indigenous materials. It's a kind of beach umbrella that looks like a house. Classic.

FUNG: It is very elegant with almost Paladian proportions.

Upon investigation, I discovered that the strongest part of the coconut palm tree trunk is found one to two feet above the ground. This determined the design, which utilizes that place for the building-load convergence. Wrapping bundles of bamboo around the coconut trunks makes the trunk stronger and less flexible.

The octagonal house plan can adapt to various uses as the need arises. Walls are of woven bamboo, which is strong, light, and allows the breezes in but keeps out direct sun and rain.

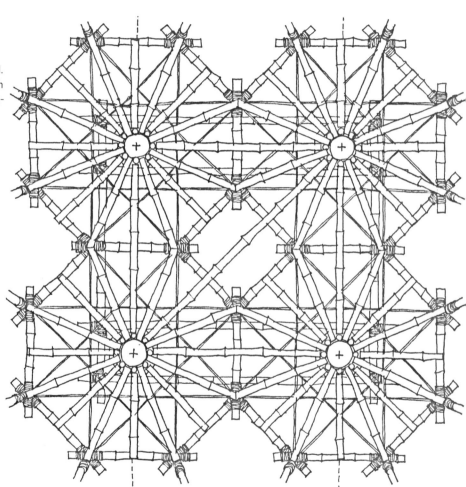

A REFUGE THAT TRANSCENDS THE LIMITS

Vision:

1. Being able to quiet the ego and join with others, while maintaining the nucleus of the individual self.
2. Modifying the sensations of "things" and increasing the intelligence of "time/space."

The outside is lived from the inside.

Eliminating divisions: An embracing cane strip, a horizontal platform, and a curtain with vertical mobility determine the limiting of spaces.

Eating, sleeping, and resting are understood as a series of related activities rather than separate functions. Circulating corridors and the spaces they join are indistinguishable, so the structure is liberated from traditional hierarchies.

There is heterogeneous mobility in a space that can be appropriated in diverse ways.

Mike Russell
17 USA

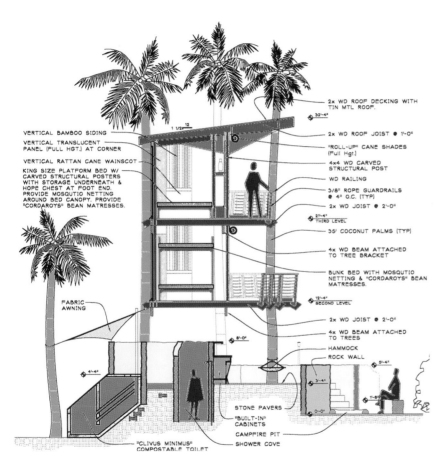

VERTICAL BAMBOO SIDING

VERTICAL TRANSLUCENT
PANEL (FULL HGT.) AT CORNER

VERTICAL RATTAN CANE WAINSCOT

KING SIZE PLATFORM BED W/
CARVED STRUCTURAL POSTERS
WITH STORAGE UNDERNEATH &
HOPE CHEST AT FOOT END.
PROVIDE MOSQUTIO NETTING
AROUND BED CANOPY. PROVIDE
"CORDAROYS" BEAN MATRESSES.

FABRIC
AWNING

2x WD ROOF DECKING WITH
TIN MTL ROOF.

32'-4"

1 1/2" 12

2x WD ROOF JOIST @ 1'-0"

"ROLL-UP" CANE SHADES
(Full Hgt.)

4x4 WD CARVED
STRUCTURAL POST

WD RAILING

5/8" ROPE GUARDRAILS
@ 4" O.C. (TYP)

2x WD JOIST @ 2'-0"

21'-4"
THIRD LEVEL

35' COCONUT PALMS (TYP)

4x WD BEAM ATTACHED
TO TREE BRACKET

BUNK BED WITH MOSQUTIO
NETTING & "CORDAROYS" BEAN
MATRESSES.

12'-4"
SECOND LEVEL

2x WD JOIST @ 2'-0"

4x WD BEAM ATTACHED
TO TREES

8'-0"

HAMMOCK

ROCK WALL

5'-4"

3'-4"

1'-8"

4'-4"

0'-0"

STONE PAVERS

"BUILT-IN"
CABINETS

CAMPFIRE PIT

SHOWER COVE

"CLIVUS MINIMUS"
COMPOSTABLE TOILET

RUSTIC RETREAT

To recapture the flavor of my childhood growing up in the mountains of western North Carolina, climbing in a tree at my grandmother's house, this treehouse is a combination of my roots and the four elements: earth, fire, water, and air.

A low noncombustible wall surrounds a campfire pit directly in front of the treehouse on ground. Thus, fire is the ground-floor element.

A short run of steps, just inside the campfire area, climbs through a narrow passage to the elevated stone-paved first floor, three feet or so above the ground floor. A hammock would be strung between two of the seven 35- to 40-foot coconut palm trees needed to build the retreat. This would be the element of earth.

Stone steps lead down to the sunken, stone-wall shower and pedestal sink basin—the element of water. A composting system would be placed at the rear of the first floor, away from the living area.

Wood spiral steps wrap a single palm to the second floor viewing deck and built-in sleeping bunks. On the fourth side of the bunks are roll-up bamboo shades. Here we have the element of air.

The spiral steps continue to the final level, the third floor, and a sleeping hideaway in the treetops some 30 feet above the ground.

The structure consists of wood beams that are tied to posts of coconut palm trees. Simple hand-carved beams of native wood would run between each pair of trees.

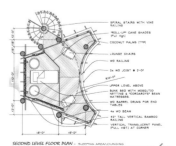

SPIRAL STAIRS WITH VINE
RAILING

ROLL-UP CANE SHADES
(FULL HGT.)

COCONUT PALMS (TYP)

LOUNGE CHAIRS

WD RAILING

2x WD JOIST @ 2'-0"

UPPER LEVEL ABOVE

BUNK BED WITH MOSQUTIO
NETTING & "CORDAROYS" BEAN
MATRESSES

WD BARREL DRUMS FOR END
TABLES

4x WD BEAM

43" TALL VERTICAL BAMBOO
RAILING

VERTICAL TRANSLUCENT
PANEL (FULL HGT.) AT CORNER

SECOND LEVEL FLOOR PLAN - SLEEPING AREA/LOUNGING
SCALE: 1/4"=1'-0"   APPROX. 194 sq.ft.
NORTH

The treehouse is more than a house built in the trees. It is at once a floating object and an open extension of tree space. The interior and exterior are never distinct, but ever changing. The treehouse builds on the existing qualities of the trees it occupies to define itself.

The structure takes on multiple perceptions through the use of movable, lightweight wall panels. These wood-framed panels covered with local translucent fabrics may be latched closed or fully opened. The remaining walls and roof structure are covered in overlapping translucent fiberglass shingles. The translucence of these materials provides a changing, filtered light, much like light perceived through the leaves of trees. With minimal connection to the ground by ship's ladder and the addition of a small bathroom and kitchen, it is possible to live solely in tree space.

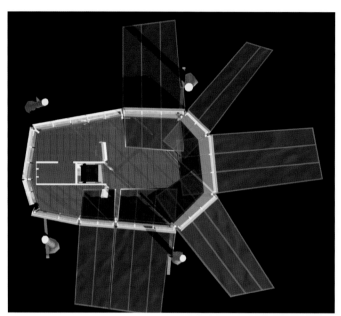

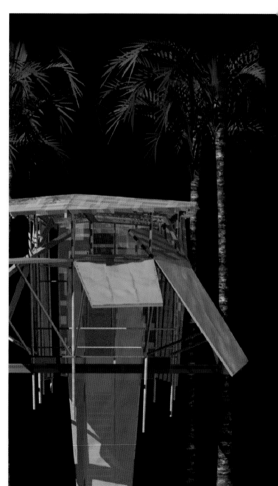

BURTON: Nice being so lightweight and looking a little like a lifeguard stand.

GREENBERG: Rather unique in its use of so many lightweight materials. It seems to me so functionally simple while simultaneously a work of art or sculpture.

GRINSTEIN: This is like a floating tent. The nice thing is the way it lets the air and the light in while still protecting you from the elements. I like the canvas.

Mark Ayers
19 USA
Top Ten Finalist

The childhood dream of a treehouse elevates the child to the height of birds, overlooking the adult domain below. As adults, we feel that a raised view can bring heightened awareness and inner harmony. A transformation occurs when our perspective is altered.

This treehouse is designed around the concept of mental and physical sanctuary. The design is rooted in Polynesian construction and spiritual consideration for body and mind. The ancient Hawaiian *hale* (home) and *heiau* (temple) were simple in form and designed for resilience and utility. The construction techniques for each hale were similar, but each was modified for its specific purpose within the group.

Our interpretation unites multiple hales into one, tied to a single vertical ridgepole or *pou-hana* (royal palm). The ground level holds the functional utilities and is very open to the surrounding environment. The first level is elevated as the social space for eating and conversation. The second level, for rest and dreamful sleep, begins to layer its façade for privacy. The top level is the most refined, and derives from early eighteenth-century etchings of ritual meetings and the surrounding spaces; it is reserved for sacred uses of meditation and personal growth exercises.

Continuing the Polynesian ways, this treehouse is designed for stewardship of the land and use of available materials. With the exception of the rigid collar connection, all parts are natural. Primary support is provided by the corner posts, or *pou-kihi*, and ridgepoles. The rigid collar anchors into the central palm and allows each floor lateral movement with the floor plates resting on top of a beam.

"Retreat . . . for a time invaluable together and alone with roots of civilization, with a people before people, with laws, *kapu*, which hold for it respect and reverence. These are the roots, the branches, the grass, the *tapa*, and the canvas for living and believing again in being human."

NOTE: *This design will be available as a pre-fabricated bamboo treehouse in the near future. For more information contact the author (see page 156).*

GREENBERG: Using the traditional thousand-year-old design features of the Hawaiian *hale*, the architect has deconstructed it in a way that is rather exciting.

What is a treehouse?

Think about being a tree, and here comes a guy who says, "I'm gonna build a house on top of you! And I might even cut you at the top, so you won't grow and upset my house."

What I'm saying is this: A treehouse is NOT an environmentally friendly project.

That's why I've made this design easy on the tree.

The way up is not easy, but it's easier than climbing, and it uses no electric power. I calculated that the force needed to raise the elevator is divided by eight when it reaches the handle of the wheel. Two simple mechanical devices would keep the elevator from going down when the wheel is turned to ascend, and adjust the pace of the descent. I propose three elevators instead of one, as it would be very difficult to raise one elevator to the third level. However, with an electric motor, one elevator would suffice.

Remo Sciessere and Marcus Weiss
21 Switzerland

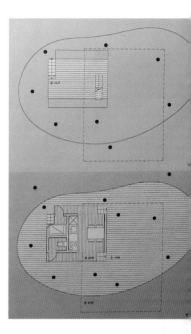

The image of a treehouse: to be on top looking down, to have unhindered vision, to have a life free of time constraints; to experience nature's rhythm as our own, to have enough time and space for timeless solitude and togetherness with one another; the search for the simple life nestled in the crown of a majestic tree.

The exterior is part of the interior living space. Our concept is analogous to sand sculptures—small forts from which one can peek and venture forth only to return and feel protected and safe.

Stairs that swing free lead to the highest platform that rests on studs attached to the palm. The sleeping quarters are surrounded by an open space, open on top. The 360-degree platform can be interrupted in order to enlarge a sleeping space. Near the stairs there is a space that encourages gathering. The room height is seven feet while the walls are only five feet tall. The roof is transparent and covered with palm fronds for shade. The treehouses stand alone, maintaining the feeling of solitude within the treehouse colony. It is essential to distribute the treehouses with reference to the topography.

Karim El Ansari
22 Germany

The secret to these structures being supported by just one tree each is in their very lightweight design. Essentially bamboo, thin cables, and ropes for access. Sizes and shapes vary a bit as they are clumped together.

GREENBERG: It is possible that a single palm tree would have a lot of trouble supporting these structures.

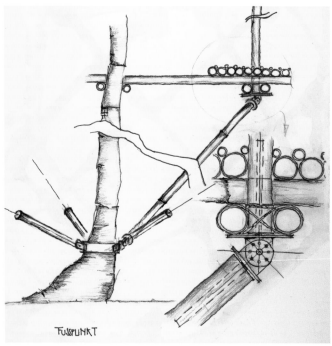

FUSSPUNKT

Vlad Andreescu
23 Romania

The treehouse platform is assembled on the palms with circular metal pieces gripped with wedges. The platform is anchored by the two palms with ¼-inch steel cables. Anchorage leads from the extreme points of the platform toward the bottom of each palm. In the same way, cables are anchored to the palms with circular pieces and wedges. The treehouse can be let down to the ground level with the easily dismounted wedge system, and in this way can be anchored strongly during a typhoon.

The structure of the house and platform is bamboo, while the partition and external walls are made of light wood. Movable frames for the walls are covered with transparent plastic. The roof structure is a bamboo network covered with a waterproof vinyl sheet; a light wood layer is added for insulation and aesthetics.

Jorg Kriebel

Ales Seligo

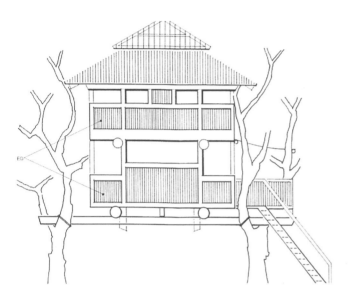

This is not a pavilion with the usual amenities for comfortable living, but rather a simple wooden treehouse with coherence in nature. A mix of one- and two-story treehouses, colored for identification, might become a small village.

The form is quadratic and much of it would be prefabricated. Toilets and showers would be housed in a separate building (a conscious choice, in order to avoid any emissions into nature).

There are monsoon forests on some Pacific islands, where nature is untouched. The shores of these islands are bare and therefore ideal for development of tourism.

We would place our wooden holiday treehouses between newly planted palm trees. Everyone would have an equal view.

Humidity demands a wooden, airy structure with sliding curtains of bamboo stalks, giving transparency to the building. A light roof made of wavy sheet metal in the color of the palm leaf would give the impression of a light pavilion raised in the air. Materials such as bamboo would be simple and natural.

Altessandro Giovanni Barilani, Valerio Tinazzi, and Lorenzo Rossi

26 Italy

Takeo Morinaga, Hiroyoshi Takiguti, and Takafumi Hidaka

27 Japan

The project evolved out of different aspects of the concept of habitation and from the theme of the competition. We were asked to project a structure not rooted in the earth, but rather a fluctuating/floating habitation in the air, anchored to some trees.

While it can move and confront the elements it must maintain strongly the basic characteristics of a domestic living space.

A large covering with a curvilinear gait to transmit the idea of shelter that wraps around you.

A platform, like a raft of timber, is anchored to the trunks of the palm trees, using garnier joints in order to guarantee the trees their mobility.

Walls and windows used as membranes, adjustable to the varying temperatures of day and night, like living walls.

Access stairway is the sole point of contact with the earth; this provides a hidden place for the infrastructure (water, electricity, gas, etc.).

It is composed of two types of zones: public in the middle, private on both sides. The central zone is not only for living, but enjoying the resort. After swimming in the sea, we can lie there on the floor and take a nap. This is the main space in the life of the treehouse.

Dave R. Miller, Ryan C. Clarke, Jaqueline Woodley, Tony Carona, and Peter Williams

28 USA

We decided to place our treehouse on Fiji's island paradise, the archipelago itself, with each island growing larger and larger as it wraps around a single point. Our hut organization is derived from this configuration. A cluster of huts face an inner court where group activities like bonfires and beach parties could be held.

We found the coconut palm ideal—strong enough for our purpose and easily transplanted. We looked to its fruit, the coconut, as the inspiration for the structure and form of the units. But while the coconut is solid, heavy, and dense, we wanted our units to be the opposite—light and airy. So we took the skeletal form of the coconut and attached a layer of fixed slatted louvers to form jalousies used for regulating ventilation in the warm, humid climate of Fiji.

We staggered each unit vertically so that each would be located at five-foot intervals from its neighbor, thus having the appearance of an ascension to, or a cascade from, heaven. Each hut would be suspended between a cluster of at least four trees, using a thin steel cable. The cable would be secured to the bark of the tree by means of a beltlike brace, which would house a type of cable raceway lead. At the center of the floor of each hut would be a "king post" where all cables at once converge and diverge, thus forming a tension loop. The system would be kept in tension by the movement in each individual tree and the weight of the hut. The occupants would always be aware of slight breezes through the swaying of the trees.

Living in a treehouse is a childhood dream. A treehouse has no solid walls or fixed structure as it needs to change with the growing tree that provides the foundation. While normal houses protect from the elements of nature, a treehouse connects with the elements.

Our treehouse is like a nest that interconnects the trunks of several palm trees in a series of platforms with different functions. The lowest platform incorporates all the utilities. The sleeping platform is in the middle, with the viewing and living platform on top. A staircase leads from one platform to another, around the central palm tree. The height of platforms can be adjusted according to tree height. The design can accommodate regional building traditions.

The platforms are simple A-frames fixed to the tree trunks with special steel sleeves, providing stability with adequate flexibility. The roof follows the A-frame shape, but is bent so that rainwater can be collected for the shower tank. The roof and platform are connected by posts. The back walls are fixed with adjustable bamboo screens.

As it is a modular system, many different configurations of treehouses are possible.

All structural elements are made of bamboo. All connections and joints can be made with ropes and fibers, so that with the exception of the stainless-steel sleeves, no screws, bolts, or nails are necessary. Materials can be grown and harvested on-site.

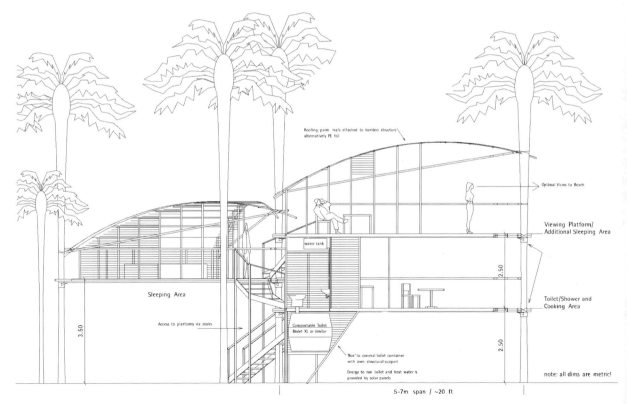

Roofing: palm leafs attached to bamboo structure
alternatively PE foil

Optimal Views to Beach

Viewing Platform/
Additional Sleeping Area

water tank

2.50

Sleeping Area

Toilet/Shower and
Cooking Area

3.50

Access to platforms via stairs

Compostable Toilet
Biolet XL or similar

2.50

'Box' to conceal toilet container
with own structural support

Energy to run toilet and heat water is
provided by solar panels

note: all dims are metric!

5-7m span / ~20 ft

This is a symbolic tree, a stem on which hangs an observation platform/housing shell, which can be imagined as a petal or a flower.

There are two main beams with double supports.

From the platform, eight stems, curved by cables, support the roofing, making a light and rigid frame.

We use a method of traditional Thai house ventilation for maximum protection from heat and sunlight.

The main frame and the floor are made of wooden boards. The roofing is made of little wooden boards or corrugated tin. Solar panels feed two fluorescent lamps.

Screens are made of plaited local vegetation. Furnishings are local handcrafted wooden table, chairs, and beds.

This place can be the stimulus for a new point of observation of one's own life, for discovering a greater relationship with the natural ambience.

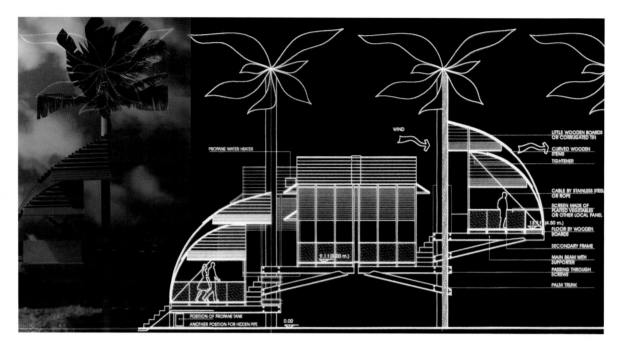

Patrizia Pederzolli
31 Austria

The tropical climate makes a very open, free design possible and there are no limits for fantasy. Where do I start to give a shape, when there are no prescriptions or instructions? I began to think about space, structures, etc., and researched natural phenomenons. My goal was to create a new, alternative world, which has nothing to do with common architecture.

I designed a project consisting of private treehouses shaped like cocoons of minimal space just for protecting the visitor from natural influences and a system of open platforms connected by paths, all suspended above ground. The base for the project is an imaginary destination characterized by the two parameters of wind and palm trees. The shape of the platforms and paths depends on the constellation of the individual palm trees. This construction principle can be used on every arbitrary constellation. To guarantee a sensitive interference with nature I chose the local material bamboo.

The concept is for a community of thirty people who are sharing eight treehouses, each for two to six people. There are ten platforms, the biggest one in the center being the only one that has a special function. From there the earth can be reached by two lifts; it is also the supply base and main meeting point. The other platforms are used individually. The paths are the connections between them and also build the base for climbing the individual treehouses by rope ladder.

My aim in design was to reach functional aesthetics through intelligent construction. The whole structure gently moves with the wind; everything is in light motion all the time. The ropes carry the weight to the palms and from there down to earth.

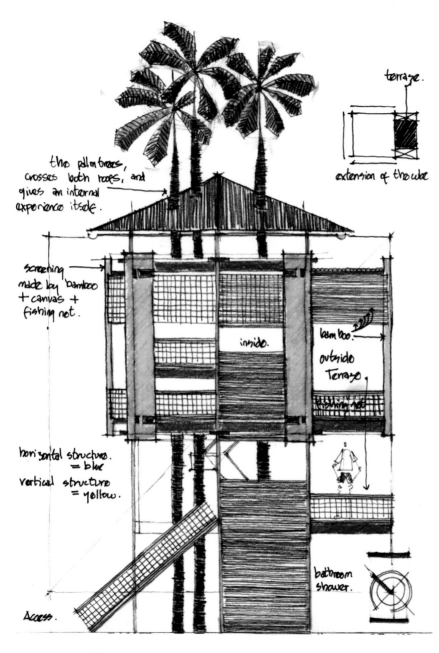

the palm trees,
crosses both roofs,
gives an internal
experience itsefe.

screening
made by bamboo
+ canvas +
fishing net.

inside.

bam boo.

outside
Terraze.

horizontal structure.
= blue
vertical structure
= yellow.

Access.

bathroom
shower.

terraze.

extension of the cube

Angelo Hubert Gadischke Fegali
32 Colombia
Top Ten Finalist

In Italo Calvino's *The Baron in the Trees*, Cosimo's hanging fountain was a kind of gutter made from the bark of a poplar tree that carried rainwater from the branches of the oak so that he could drink and wash himself. It was thought that his feelings and instincts were different from those of other men, and that the skins he had adopted as clothes represented a total mutation of his nature. It seemed that his continuous contact with the bark of trees, the movement of feathers, and the green current that travels as blood through the veins of the leaves—all these life forms—had modeled his soul.

GREENBERG: This finalist probably did the most homework of anybody in his desire to understand the simple systems that allow a structure to actually work in nature. I suspect that he also has had a lot of experience in the tropical jungles of South America.

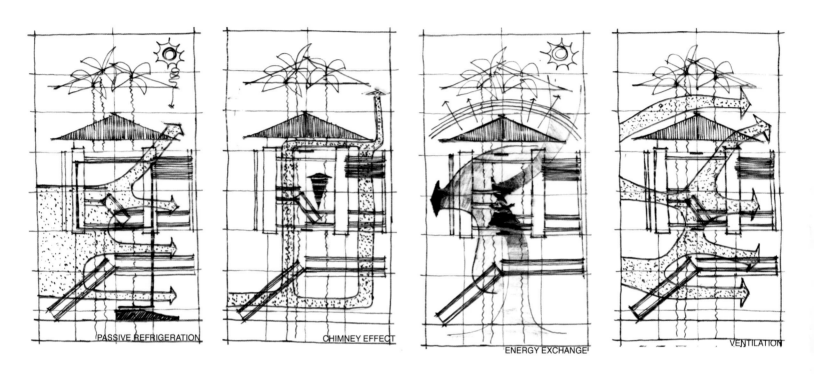

PASSIVE REFRIGERATION

CHIMNEY EFFECT

ENERGY EXCHANGE

VENTILATION

In Emptiness
Arises this Form,
Beautiful, wondrous, graceful,
Creating space for soul and silence,
breathing Emptiness:
be this Emptiness.
   — Master Chang

GREENBERG: Perhaps one
of the most Zen entries, it
couldn't be simpler. An elliptical
within an elliptical "room"
suspended above an ellipse
on the ground containing the
bathroom and kitchen. All
very lightweight, all made of
bamboo. It could use some
designation as to how it would
be attached to the trees but
that would make it less Zen!

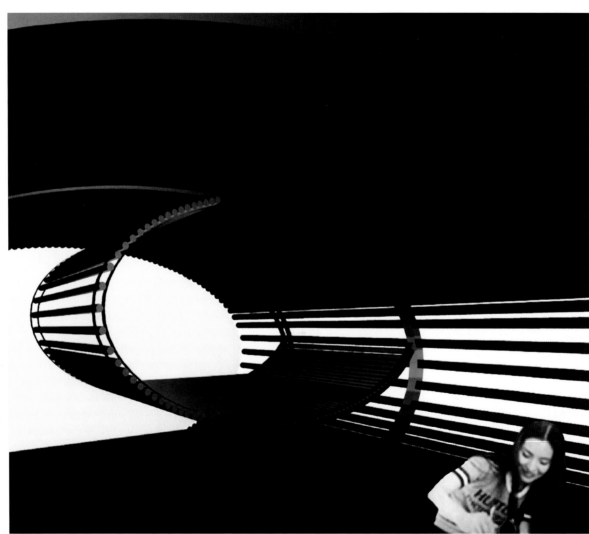

Susanne Heiss and Barbara Horst

34 Germany

This design combines two entirely different kinds of structures. The common area is a spacious and comfortable room overlooking the sea, and the private resting-places are three cocoons mounted in the trees. The open space mirrors the vastness of the sea and beach, while the cocoons shape the seclusion of life in the trees. This is similar to the arrangement found in campsites.

The party space is a platform of sixty square meters attached between palm trees at a height of two-and-a-half meters. A combined kitchen and sanitary block is the only fixed furniture on the dune side, and can be closed for weather protection. Two sail cloths are stretched over the area. The three sleeping capsules are suspended above and next to the platform at a height up to seven meters, and can be reached by rope stairs. The seaside entrance is equipped with a folding mechanism, like a ramp.

The platform is a rigid plane mounted to the palm trees on flexible connections. A layer of bamboo strips is fixed to wooden beams of local wood. Recycled oil tanks contain water supply and sewage system; electricity supply is secured by solar panels. The cocoons are constructed with a mesh of split bamboo cane (mesh width 35–40 cm) curved into a three-dimensional grid shell. The bamboo grid is covered with a translucent cloth.

Vitoriano Hernandez and Jose Minguell

35 USA

The Casabaza Treehouse consists of three baza-shells and three open baza-platforms. The baza-shells are the supporting framework for activities. An internal, removable "banana skin" can be zipped to the wooden structure for privacy and foul weather conditions. The open spaces serve as a leisure zone, protected by canopies. A net is stretched between the platforms, in which to meditate while floating weightless above the ground. Or, the net can provide storage space. The baza-shells are suspended by cables from three palm trees which are placed in an equilateral triangle. The units are local hardwood. The platform, acting as a compression ring, is constructed of local materials. The canopies and internal skin of the modules are fabricated by local craftsmen. The tension cables are steel or structurally equivalent fiber. A suspension system was selected in deference to the natural growth of the palm tree; it allows for prefabrication.

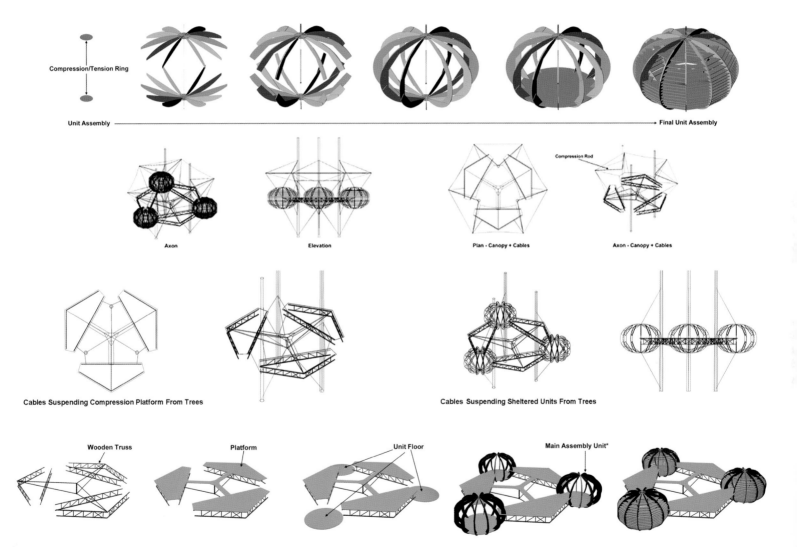

Compression/Tension Ring

Unit Assembly ———————————————————————————————————→ Final Unit Assembly

Axon

Elevation

Compression Rod

Plan - Canopy + Cables

Axon - Canopy + Cables

Cables Suspending Compression Platform From Trees

Cables Suspending Sheltered Units From Trees

Wooden Truss

Platform

Unit Floor

Main Assembly Unit*

Platform Assembly ———————————————————————————————————→ Final Platform Assembly Unit

1st deck

2nd deck

luggage
lift-up
system

ladder

terrace

SUSPENDED TREEHOUSE

The treehouse is conceived as a cradle suspended in the South China forest. This precisely balanced bamboo structure houses essential functions on different levels to maximize the openness and the experience of progressing through the space. The design is inspired in part by the elegant functionality and structural integrity of an insect's nest.

The treehouse is connected to, yet independent of, the adjacent trees and their natural movements, allowing for flexibility and variation in site conditions. The design neither interferes with the growth of the trees nor imposes a heavy load above their bases. The bamboo framework and network of support cabling allow air to flow through the structure, further reducing the impact of extreme wind loads. Bowed bamboo supports bear the required stress while contributing minimally to the overall weight of the structure.

A visitor enters the treehouse via a lower-level terrace accessed by a retractable ladder. A basket can be raised from the ground level to the terrace to transport luggage and other heavy items. A two-flight staircase leads to the first deck containing bedrooms and enclosed bathroom areas with chemical toilets. The visitor progresses to the second deck, which functions as a kitchen, dining, and storage space, and finally to the third and topmost deck, which includes a master bedroom and bathrooms and is oriented toward the ocean.

In response to its context, this design incorporates concepts of green architecture and social responsibility. Natural indigenous materials are used, including bamboo for most structural elements and palm leaves for the roofing. No electric power is required for building the units and domestic, low-skill labor can be employed for construction.

GREENBERG: The design is well thought out in its use of bamboo.

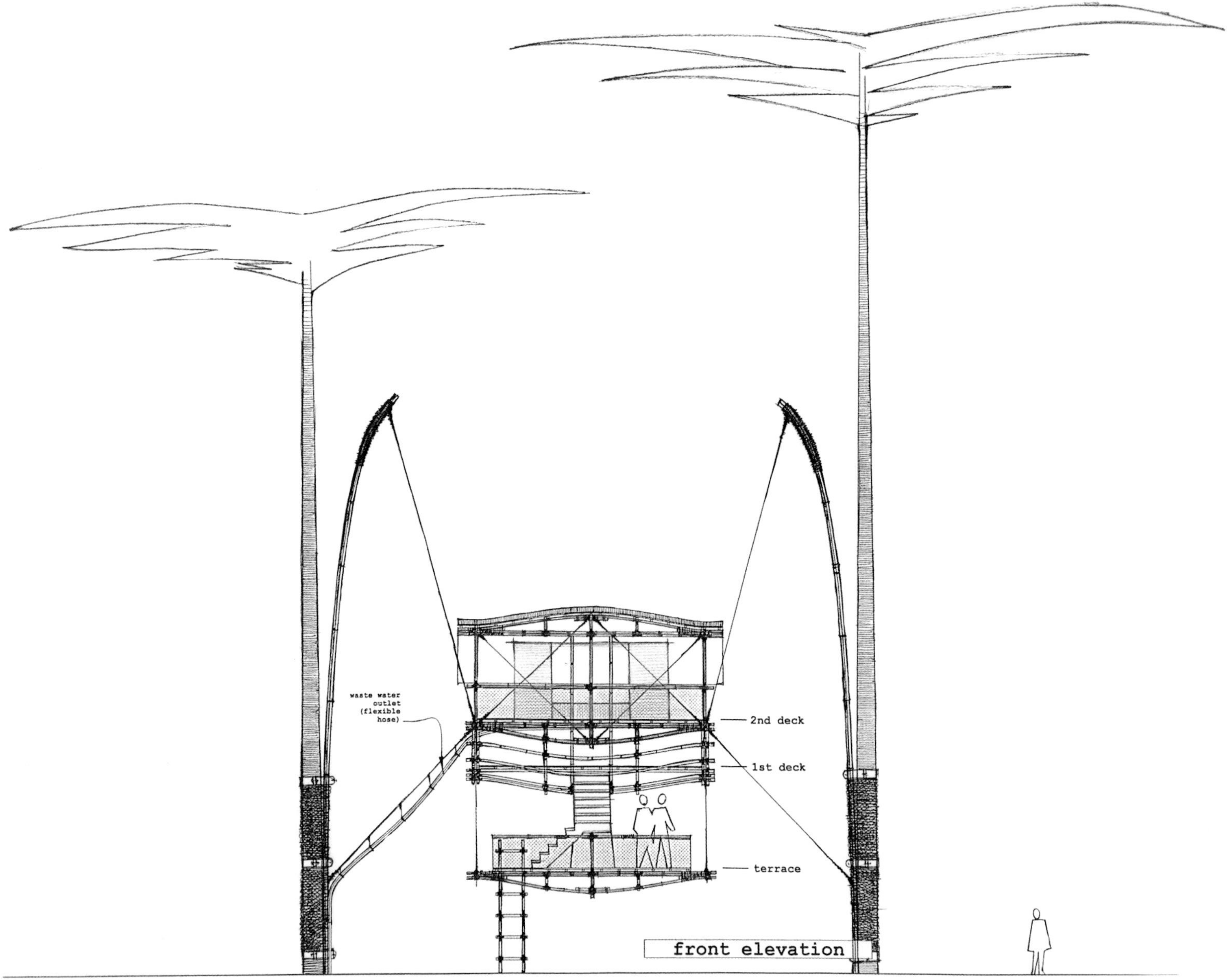

waste water
outlet
(flexible
hose)

2nd deck

1st deck

terrace

front elevation

First treehouse designed and built by the author, front view, in Hana, Maui, Hawaii, 1996.

Introduction
David Greenberg

II. MAKING A TREEHOUSE

On New Year's Day 1996 I began to design a treehouse for my property, a 20-acre plot of land at jungle's edge in Hana, on the island of Maui, Hawaii. It was one of the few structures I did not learn about in architecture school.

A good treehouse is a rather anti-architectural statement. As I have always been a bit of an anti-architect anyway, I began the project with great relish. I walked up into the jungle of Hana where a big old kukui-nut tree was standing and started sketching some designs. There was already a little platform in the tree where some kids had built a fort a long time ago. It wasn't easy to figure out what to do. I drew the existing tree and platform and started to draw a structure attached to it but nothing seemed right. That night I spent hours with tracing paper over my sketch of the tree, but nothing took shape. The next day I took photos of the tree, had them blown up, and put tracing paper over the photos, but again, nothing. This had never happened to me before. My designs had always come to me so effortlessly, almost like magic. But this time no magic, no nothing.

I struggled with my designs for three weeks with no success whatsoever. Then it hit me—I wasn't going to draw the design, I was going to just do it. This was a highly unusual plan of action—I had gone to architecture school for over ten years, including a couple of years in graduate school for urban design, and I had taught architecture for another ten years at UCLA (ext.).

Author's first treehouse, middle level.

The last thing in the world I would have thought of doing was to build something without a plan. But this was different. In this case, I had reached the point where I felt that the only possible way of building a proper treehouse was to, in fact, not design it, but to let the tree design itself. It was a counterintuitive method for me, but at the time it felt like the right thing to do, so I just let go of my doubts and questions and went with it.

I began work in mid-January 1996, with two assistants. They helped a few hours a day in exchange for staying in a cottage of mine. One was a carpenter and the other, from Guyana, had grown up with a machete in his hand. Their names were Jay and Joel, respectively. Jay had a lot of tools, Joel had only a machete. But as the site was in the middle of a jungle overrun with guava, the machete proved indispensible. You know the expression "shit happens"? In Hana the expression is "guava happens."

To begin, Jay and I found a two-by-four in a pile of discarded wood. It was sixteen feet long and just barely spanned the distance from one kukui tree to another. We put it somewhat arbitrarily at ten feet high. It became the end joist for the main floor of the treehouse. My next move wasn't immediately clear to me. I looked around but none of the guava trees was close enough to reach with another two-by-four, so I yelled over at Joel: "Hey Joel, bring that tree over here." The tree was over twenty feet high with a diameter of six inches. Joel cut down the tree at its base, and after trimming it a bit we placed it on top of a lava rock for footing, thereby creating an equilateral triangular floor plan of about fifteen feet on each side.

Author's first treehouse.

In one hour we had created the basic floor plan that fifty hours of attempts at design hadn't accomplished. The act of doing it had begun the "design," and from then on it just happened, like guava happens. Jay started filling in the triangle with two-by-fours, about two feet on center. Joel brought other trees to us that supported the fifteen-foot span midway and were tall enough to go up another ten feet to support the roof of the second floor. While Jay and Joel were busy relocating trees, I was trying to think of what to do next when I remembered that I had once been inside a truly great treehouse.

Deep Background

About twenty-five years ago I was visiting the Island of Kauai and picking mushrooms in the middle of a cow pasture, when it began raining. I found cover under a thick grove of trees. In the grove I found a couple of guys who had already picked their limit of 'shrooms. When I told them I was in graduate school studying architecture at UCLA, they insisted I come and see the house they had built where the jungle met the beach. It was located on a piece of property owned by Howard Taylor, as in Elizabeth's brother.

This area was called Taylor's Camp and evidently he had given permission for about fifty hippies to live in the trees on the beach. They had built about twelve treehouses and they were all spectacular. I went from one to another in total awe and disbelief. As it happened, I was interested in alternative architectural solutions at the time, and they had used a lot of bamboo and a clearish vinyl for the roofs to keep the rain out. We ended

Guanyin View treehouse, designed by the author, at Nanshan Buddhist Culture Zone, Hainan, China.

up in the nicest of the treehouses, climbing up a bamboo ladder to a room filled with throw pillows covered in Hawaiian-patterned cloth and a grass-mat floor with a moiré pattern of leaf shadows created by the sun filtering through the trees. I reclined in the pillows as my hosts began to play some musical instruments. I became mesmerized by the shadows of the leaves moving on the floor. I'm sure the 'shrooms had something to do with the total enjoyment of the environment, but I had finally found as perfect a design for a structure and interior as any I had yet seen or experienced. After a while they all decided to go for a swim in the ocean across the little sand beach. I stayed behind to rest. I began to hear the wind blowing through the trees. The sound of rustling leaves became louder. The last thought I had before sleep was, "What beauty." I awoke to the sound of a big wave crashing on the beach and went down to bodysurf.

When I finally found the picture I took of that treehouse I slept in so many years ago, I was amazed at how beautiful it was. I also knew I had to find some bamboo! Fortunately, there's a lot on this side of Maui, but many of the forests are remote and hard to get to. And commercially it's two dollars per foot; I needed a lot, and I had little money. A few days later I got lucky. I had put the word out on the "coconut wireless," the main form of communication in Hawaii, that I needed bamboo. Turns out a friend of a friend had a lot of thick golden bamboo (the kind with intermittent green stripes) that was threatening the phone and power wires on his flower farm and was looking for someone to get rid of it for him. The next day I was there bright and early with a chain saw. It was

Gao Xiao Lan, CCTV journalist, resting before her interview with the author on the deck of Guanyin View.

The stairway in Guanyin View.

quite a task. The winds were gusting and blowing the bamboo against the wires. We had to tie ropes around bunches of it high up, and then saw and pull just as the wind blew the bamboo in the right direction. We cut fifty stalks in this way over two days at an average length of twenty-six feet. We had to build a special wood frame on my vehicle to haul it all back to my farm.

I now had building materials, an unlimited amount of guava wood, and a shitload of bamboo. Every day I spent an hour or so at the lovely Hana dump (looks like a country club) looking for odds and ends of plywood and two-by-fours for the decking. The treehouse had gone pretty far in designing itself, but it was now time for me to take charge. I had found some flexible black vinyl cords, a garbage bag full of about a thousand pieces two feet long, which had been made for use in irrigation. It was perfect for tying the bamboo into place; quick and easy, temporary or permanent, as well as handsome. It felt like I was "sketching" the rest of the treehouse with the bamboo itself, at actual scale, in place. The overall form would be finished in about two days.

Finally, the skeleton structure was up. It was exciting. I called Major Ed James, a specially trained psychic for the U.S. government, who had envisioned my treehouse even before I had, to tell him of my success and kidded him that I was taking reservations for the treehouse for Easter vacation. I had been working feverishly, getting up at the crack of dawn and working until it got dark, but in fact I didn't really know what was next. That next morning I heard the sound of the fax receiving something. It was a poem from one of

Master Chang meditating near Hana, Maui, Hawaii.

Ed's friends in Century City, California:

TREE HOUSE ME
I long to be aloft in a tree
The limbs of an elm gently cradling me.

To ponder, stargaze, or perhaps just to dream
Of quest and journeys as yet unforeseen.

Realities shift and visions unfold
Of what can be and is yet untold.

Its name shall be Ashtabula's Arms,
Majestic and bold with infinite charms.

Color me there in spirit and soul
Until I attain my dream; this goal.

—Link Schwartz

Exactly what I needed, a poem about a treehouse. Not much of a help. But when I went to the tree site that day it looked different. It was alive, to be revered and honored, no longer just a structural element, but a beautiful living thing, something of joy. Despite myself, the message of the poem had imperceptibly changed much in my mind. I now had a more feminine and spiritual viewpoint, one that was changing my mood. I was now challenged and guided by something very different. Each limb of the tree took on a new meaning to be dealt with both aesthetically and functionally. I had thought to build a stairway to the third level; the poem led me to use the existing branches of the tree as a spiral stairway, and it worked beautifully. I had previously planned to enclose most of the third-level sleeping area except for a couple of

The author and a friend in hammocks on the top level of Big Beach in the Sky treehouse, Hainan, China.

windows; now I decided to have lots of windows so that you could see and feel the tree all around you, especially the pretty gnarls in the limbs. I created a whole other space on the main floor, where you have to crawl under or over a major limb to move around in the space. I was allowing the shape of the tree to direct the design rather than my own more linear thought process. The space took on an even greater meaning for the uniqueness of its entry—which could only be accessed by crawling, bowing, climbing, or all of the above—as well as the feeling of the tree surrounding and embracing you.

The poems continued to arrive by fax every few days, and they became the basis for the spiritual plan that would imbue the treehouse with its greatest spirituality. Though the tree didn't become more anthropomorphic particularly, it did become more of an equal, a friend, to me. Sometimes I would be up on the deck working and thinking about the next move or detail when one of the lines of the poem spoke the move to me, and it was perfect, and I would jump up and down for joy. I felt a little closer to heaven with each jump.

Anti-Architecture

I've always been a bit of an anti-architect. Even as a student I rebelled against the traditional ideas of modern architecture in the 1960s. It was partly from this experience that a sort of epiphany occurred, that an idea began to take shape. I came up with a name for this abstract notion of mine late one night during a UCLA seminar I took in 1972 called Man-Environment

The beach that embodies the author's Big Beach in the Sky mentality. Sanya, Hainan, China.

Relations. We met the professor far out in Death Valley, and we stayed up all night under the stars discussing, debating, arguing. A lot of the other students were admirers of the newly built World Trade Center and the type of urban design that's rampant in this country today—bigger was better in their minds, forget the environment. But not for me. That night I found a name for my idea, and I called it the Big Beach in the Sky. I decided that night that this was to be my path. I was not going to be a typical urban designer. Big Beach in the Sky is my name for a kind of paradise—it's about nature, it's about respecting the environment, it's about humans living in harmony with their surroundings. To me this is heaven. And it is this way of looking at things that ultimately led me to treehouses. The first treehouse I built in China, the one you see in the photographs at the beginning and end of this section, I call Big Beach in the Sky.

I had spent twenty-five years talking about anti-architecture. I was now having a great deal of fun practicing it with my experimental treehouse, especially in my hunt for materials. I went to the dump every day, looking for scraps of wood or anything. Anything that looked remotely useful was thrown in the back of my jeep and hauled away. I picked up rocks from the ocean for paths. The rocks chosen were always the largest and flattest, the biggest I could possibly lift. Anything free was a potential building material. I would spend time each week also going through the back alleys of the industrial park of Maui looking through the Dumpsters. I found ten screen doors, a major score as the main sleeping loft now became almost entirely screened in.

Night shot of Treetops treehouse, designed by
the author, Hana, Maui. Photo by Peter Guttman.

My favorite material was guava wood. I greatly enjoyed cutting the trees down and carefully cutting the limbs off to be used for balustrades or even major supports. It was a real test of my brute strength. Nothing challenges as many muscles as the work of a lumberjack. At one point I bought a chain saw, but it only worked half the time, used too much gas, and I ended up spending hours sharpening it due to the hardness of the guava. Besides, the noise, the oil, and the smell didn't seem to mesh with my newfound spiritual side. My sister sent me a big bow saw from Canada, which allowed me to work more smoothly with the guava.

One day I twisted my ankle so badly that I was on crutches for a week and walked with a limp for a month. In some ways it was a good thing. In my eagerness, I had been working much too fast, and now I was forced to slow down. It is very difficult for me to sit still, but while I lay or sat with my foot up for hours at a time I began to read a book I had discovered in my meager library: *The Way of Zen* by Alan Watts, published over thirty years ago. I flipped through it and stopped at the last chapter, "Zen and Art." I got halfway through the chapter when I noticed the sun was setting, so I hobbled out to the balcony where I watch the view at that time of day. It had been a grayish day so it was not going to be a great sunset. It was quiet. It's almost always quiet except for the sound of a few birds and sometimes the whisper of faraway waves. The sounds of the birds gave me the idea to whistle. I wasn't very good, but I knew I had an hour or so out there so I took my time. A yellow-beaked minor bird came to sit on a branch just a few feet

Big Beach in the Sky treehouse.

away. I began to whistle to that bird, really getting into it. My attempts got a response as the bird hopped from one branch to another coming ever closer. After a bit, another bird joined in and then another. I ended up giving a concert to about five birds! I went back inside, flopping (carefully) down on my bed and picked the book up again. The very next paragraph I read essentially described the experience I had just had. It is called "wabi," which is described by Watts as the kind of Zen experience of enlightenment brought about when an artist gets a little depressed and has a very simple experience in nature. Following "wabi," he says, is "sabi," the next level of enlightenment. I began to understand the art of Zen. Actually, I already half appreciated the world in a Zen sort of way but hadn't realized it.

Then I remembered that an old friend, Leonard Koren, had sent me a book he had written called *Wabi Sabi*. I found it a month later, and read it with relish. It explains a simpler way of looking at art and nature and how they affect each other. Maui had taught me well over the years to see art in the aging of things. I saw the simple beauty of time in the Japanese river-washed stones everywhere, from under the sea to the top of Haleakela Crater. I remember I was driving once along the road to Hana and upon turning a curve in the jungle was flabbergasted to discover this valley filled with about twenty-thousand shades of green. In college art classes we had identified about twenty. And there were the twisting red rivers of molten lava—like kinetic sculpture on the Big Island of Hawaii. I was once on a deserted peninsula at the southwest end of Maui and

Interior of Hawaiian Hale treehouse at Nanshan Buddhist Culture Zone in Hainan, China. Design by Francis Sinenci.

happened upon a little cove filled with broken pieces of coral that had been smoothed by the water over time. Mounds of them glistened in the sun, and I thought they looked like rare gems. They were dazzling.

Reading further in the chapter I got to the part about the apotheosis of most forms of Zen art, the classical Japanese teahouse. Though I had been in a few fine ones, his description helped me better understand their hidden meanings. There is the simple architecture in its enclosure, works of art on the wall, and a flower arrangement, but best of all and most simple are the teacups. Selected by great Zen masters after examining thousands, perfect simple cups are glazed only on the inside with just a few small drips around the edge, the perfect accidents. The server offers the guest of honor the best one with the drip facing him. The more I understood about teahouses, the more I thought about building one, and I still hadn't finished the treehouse.

Another vivid experience happened to me a few years later, and it led me to yet another spiritual epiphany. Close to five-hundred business men in dark suits were gathered in a hotel ballroom at the behest of the White House and the United States Embassy in Hanoi. I was one of them, and I was the only one wearing a Hawaiian shirt. Many had arrived on Air Force One . . . I had come by boat. The time was November 2000, and we were at a joint signing ceremony during President Clinton's only official visit to Vietnam. Along with representatives of seven major U. S. companies, I stood in line at the front of the hall

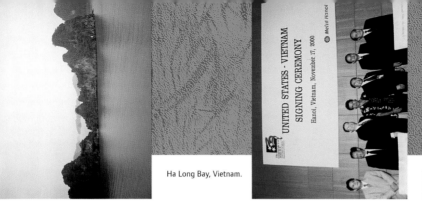

Ha Long Bay, Vietnam.

The author (center, in Hawaiian shirt) with, from far right: Vietnam Minister of Commerce; author's client, Mr. Dong, of Huong Giang Tourism Co. of Hue, Vietnam; U.S. Secretary of Commerce, Norman Mineta.

where we would all sign contracts. Before me in line was the Asian regional chief of Boeing. He was selling three 777's to Vietnam Airlines, a deal worth more than $1 billion. Behind me was the president of Oracle, who was supplying the entire country with a new software infrastructure. When it was my turn to sign the papers, I half-jokingly apologized to Norman Mineta, the U.S. Secretary of Commerce at the time, for having such a small project (my company, Treehouses of Hawaii, was to build a treehouse resort in collaboration with the governor of Hue province). He smiled and assured me that it must be a good one if his staff and the president had chosen it. (Rumor has it that the Secretary laughed for half an hour when he got back to the embassy and learned that my deal was valued at $100,000.)

Days later in the Hong Kong airport I bumped into the head of Boeing. He told me it had taken him almost five years to get his contract signed. I think it took me about five days. I must admit, it's still somewhat of a mystery to me as to how it was decided that my treehouses would be important to the growth of the new Vietnam, but someone, obviously, did.

When I arrived back in Hawaii, I got a call from Ms. Than Lo, president of the Vietnam–American Chamber of Commerce in Honolulu. Ms. Lo had helped me a great deal in Vietnam and was herself from Hue province. She invited me to Thanksgiving dinner the next day. I showed up with an article from that day's New York Times about my newly signed agreement to build treehouses in Vietnam and the treehouse competition we had just launched a couple of months earlier in

Hawaiian Hale treehouse, Hainan, China.
Design by Francis Sinenci.

September. It was a good day. Naturally, I ate too much of my Thanksgiving dinner, and, still jet-lagged, I was soon fast asleep on the couch.

I awoke to the sight of a Zen Buddhist monk sitting at a nearby table with my article in front of him. He was staring at me, or perhaps meditating, or both. He said, "I have been waiting for you to wake up." Though still only half asleep, I somehow knew that he meant that both literally and figuratively. He told me that he was Than Lo's brother (whom I came to know as Master Chang) and that he wanted to help me help his people. It was then that I knew I wanted to learn how to design treehouses that were Zen. Master Chang helped show me the way.

Details

Details were never particularly my strong suit. I have always been more interested in the abstract creative concepts of architecture. In a sense, dealing with a treehouse is almost entirely about the details. Every treehouse is unique, with its own lines and connections. They are handmade and take time. Once I spent half a day carving a dent in a large guava post with a Sawzall to get another piece of guava around it. I discovered the best details by accident. The little "island" protruding slightly above the main deck is there because a big branch the deck is resting on has a large gnarly elbow I decided not to cut away. This detail became precious as the tree so beautifully expressed itself in yet another dimension. Another detail was created when I gave a column a very natural "capitol" by leaving just the beginning of the tree

The author relaxing in his first treehouse, Hana, Maui. Photo by Dena Goad.

OPPOSITE: Big Beach in the Sky treehouse, Hainan, China. Photo by Shi Huai Xun.

branches at the top of a straight piece. Some details are purely visual while others are more functional, like the guava posts resting on large lava rocks on the ground so the earth won't rot them.

Some of the poems that were faxed during construction spoke of the parts of the tree in such loving terms, I now deify the tree as I do all of nature. The nights I've slept in my treehouse are perhaps the most memorable of my life. As moonlight filters through the leaves and branches, the house in the tree seems to light up like a candlelit Christmas tree. And I am the present. The sun will rise in the morning over the ocean, and its rays will penetrate through the branches and touch my eyelids, waking me up. Ah, this is paradise.

Javier Garcia-Arguelles Suarez and Fernando González
37 Spain

Three vertical palm trees are needed to support the triangular house of a little more than 16 feet to each side. Main beams are tied to the palms with metallic clips. Cantilevered platforms placed on the beams form the floor of the cabin. Steel cables tighten the cantilevers and support the cover. The roof is made of shiplap planks covered with tarpaper. When the structure is in place, cords are hung to hold the stairs.

GREENBERG: This model is made of balsa wood, but it seems that the treehouse would also have to be of balsa wood for those three trees to support the structure.

Santiago Badillo, Martin Fernandes Ares, Agustina Martire, Alfornso Paintini, and Maximiliano Salomon

38 Netherlands/Argentina

WOOD + CLOTH (adaptive-changeable-colorful) = SHELTER

The liberation of behavior requires a labyrinthine social space, one that is continuously subject to modification. The sense of getting lost gives way to the sense of discovering previously unknown paths. Only chance events occur, and general categorical laws are unknown. Effort is rewarded with perception of the elements that determine the situation. Static construction is incompatible with changes of behavior, and playfulness will lead inevitably to a dynamic space.

The first step is an extensive mapping of the vegetation. The trees are the squares in the board game. The pieces are cyclical but asymmetrical, and they are repeated irregularly. The way is relational, innocent, and unfinished. The environment interrupts and changes. The path leaves imbedded traces.

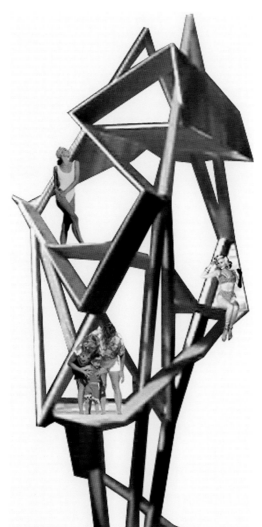

GREENBERG: The basic shapes were inspired by the stilted poles used by spear fishermen in the South Pacific.

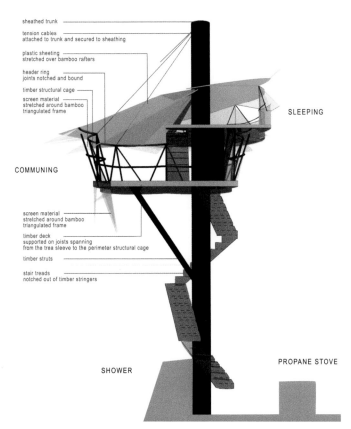

sheathed trunk

tension cables
attached to trunk and secured to sheathing

plastic sheeting
stretched over bamboo rafters

header ring
joints notched and bound

timber structural cage

screen material
stretched around bamboo
triangulated frame

SLEEPING

COMMUNING

screen material
stretched around bamboo
triangulated frame

timber deck
supported on joists spanning
from the trea sleeve to the perimeter structural cage

timber struts

stair treads
notched out of timber stringers

SHOWER

PROPANE STOVE

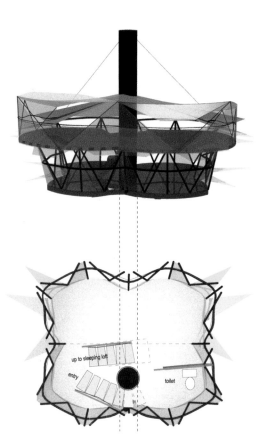

up to sleeping loft

entry

toilet

This treehouse proposal comprises a simple two-story structure hung from a sheathed coconut palm tree trunk. Water tanks and a second-story deck balance the cantilevered structure, a structural wall lattice, and diaphragm deck. Some of the structural qualities and material properties were influenced by yacht design and technology. As a collector or deflector of light and breeze, the solution provides a spacious, attenuated environment within the canopy of the resort. The skin, a combination of deployable stretched polyethylene panes, is framed in bamboo and provides a matrix of openings to regulate internal climate, light levels, and privacy. Color tints in the polyethylene skins will allow for regulation of solar radiation into and reradiation out of the building, further regulating the internal climate and temperature.

A helical stairway hung from the upper deck enables access to the lofted structure. Cooking and washing facilities are found at the tree base where a massive pedestal provides heated surfaces.

This 500-square-foot house can be placed at any of the competition sites on the beaches of Fiji, Vietnam, China, or Maui. Four coconut palms per house will be brought to the site and transplanted according to the design.

Flexibility of the interior and interconnected units is an important consideration. A "stair-bridge" accesses the house, creating a transition between land and structure. The house is divided into three principal spaces: a central space for bathroom, kitchen, and storage areas; and two flexible spaces at each side, connected by a "terrace-passage." The interior-exterior character of these spaces is created with sliding screen doors.

In accord with ecological guidelines, the building is oriented with its long axis running east-west. Rainwater from the roof is collected and stored in two cisterns in the attic; one is used for hot water heated by a solar heart located on the roof; the other stores water for various household uses. A composting toilet for four people is specified.

Although the drawings show a dimensional lumber structure, all components can be replaced with materials available on-site. The main four beams are attached to the tree with a stainless-steel collar, and the components of the basic grid are supported by or connected to the beams. The floor and roof are made of shiplap planks and the roof is covered with tarpaper. Walls are made of wooden screens, and the stair-bridge is a light wooden structure operated manually by a rope and pulley.

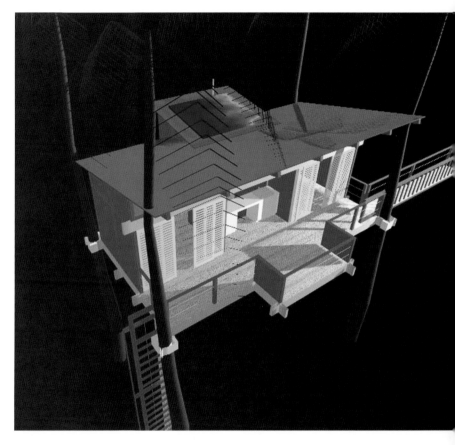

Drechsel Fromme Architekten

Dieter Fromme, Andreas Menhard, Bernhard Drechsel
41 Germany
Top Ten Finalist

THINKING ABOUT A TREEHOUSE

Wilderness/Civilization
Nature/Structure
Emotion/Reason

How does one sketch a house plan apart from a plot of land, in a perpetually changing environment? Is there a way to perform this task within the discipline of architecture? With multiple unknown determinants, a predetermined, formalistic concept won't do. The solution must be indistinct enough to contend with the uncertainties. An abstract method is required, one that is supported by a rational attitude, unimpaired by emotionally held preconceptions. The structure must invite the creative act.

As the uses of the individual units will be highly diverse, there can be no universal design. Instead, there is a persistent structuring of unlimited potentialities, allowing for variation based on reason— that is, based on the given premises. Our proposal is the notation of a general scheme for an on-the-spot realization process. This process generates a uniqueness that rejects contextualism.

What are the axioms of this method? There are no formal specifications for the form of the house, the room distribution, or the evolution of the structure. However, available regional materials must be used for fusion with the environment, and the house must be able to grow or shrink in response to the underlying principle of uncertainty in the environment.

Our structure is not an arti-fact. We want it to be the fleeting expression of the parameters of the moment. It is not a sculpture, and it is not the manifestation of an object. Thus, we have architecture that is not art.

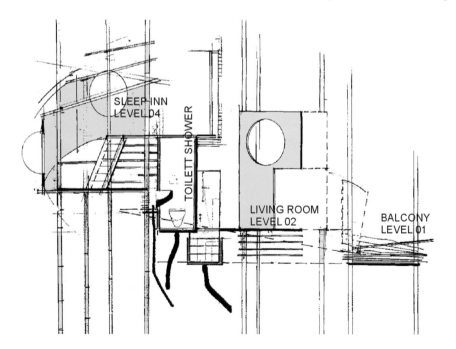

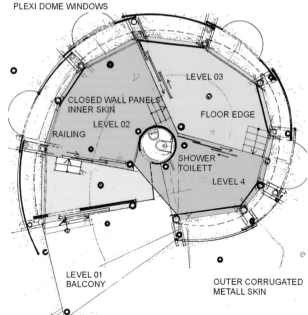

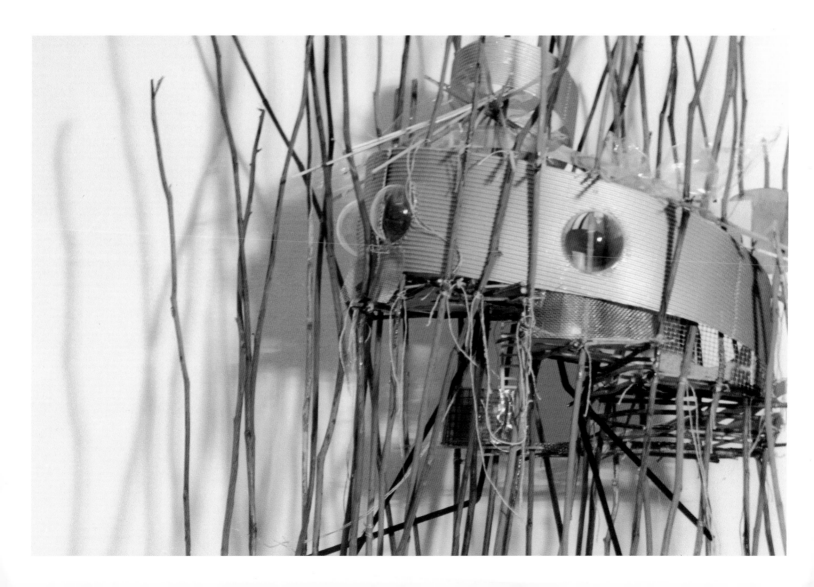

FUNG: Delightful. Very creative the way the building is not seen as much as the bamboo. Building disappears behind, deconstructing the building. We liked the idea of the platforms spiraling up to the bedroom.

GREENBERG: No predetermined exact scheme, no predetermined aesthetics—I love this. Fabulous. This definitely is a creative genius at work. A kind of Kandinsky floor plan, composition. What if? Structure made of giant bamboo. Crouching tiger, hidden dragon treehouse giving you a sense of buoyancy and strength of the bamboo and recycled materials. They also have plans showing feelings, sounds, and experiences as well as functions. A three-dimensional multisensory environment. I love the composting toilet up there. If I were the Road Warrior, I would stay in this one.

GRINSTEIN: It looks so tenuous, so helter-skelter. I love it .

The primary tree is the coconut palm and the primary building material is bamboo. Because of the climate of the Fiji Islands, the huts will not be directly attached to the palm trees; instead, they will be suspended by rope between the trees, allowing the structure to move freely and withstand high winds.

Six huts will each have three bedrooms, a bathroom, a kitchen, and an observation platform. Wooden bridges and hemp rope link the huts.

THE STATIC SYSTEM

A main frame suspended by ropes and attached to the trees will hold the entire weight of the hut. The roof, flexible but strong, is attached to the top of the frame. The ropes that attach the floor to the frame are kept taut by the weight of the floor, while forming the static structure for the walls. To minimize swaying effects, the walls are not parallel. The bathroom will have a simple shower, a composting toilet, and solar panels to heat water.

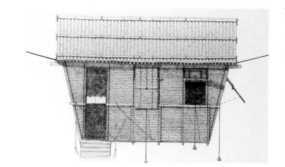

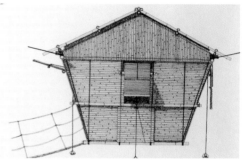

The house is a basic unit that hangs on trees. A terrace made of plaited bamboo and net extends around the basic unit. The floor and walls can be customized in various ways. The house can be easily adjusted for various uses and different surroundings. This design allows a strong connection between humans and nature. The basic structural design is copied from a spider's web—spider LOVE!

GREENBERG: Though many of the designs integrate structures from nature, the concept of the spiderweb is particularly creative.

1. Bamboo stair
2. Kitchen
3. Shower
4. Sleep
5. Container
6. Eat
7. Bamboo terrace
8. Net
9. Palm tree
10. Steel cable

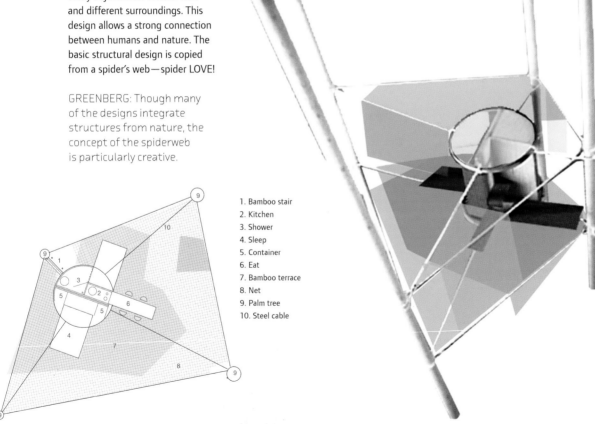

Tom McInerney
44 USA

While researching the nature of the treehouse, I read the excellent adventure book *The Baron in the Trees* by Italo Calvino. The main character, Cosimo, commits his life to a unique exile in the trees; the book tells of his quest to never touch the ground again. Yet the story was fiction. Trees sway and bend in the wind. The biggest challenge in designing a treehouse was how it could communicate stability. How can one feel grounded to the earth yet still remain aloft among the branches?

My concept begins with a pentagon symbolizing strength and security and to the pentagon by placing a tree at each vertex.

The treehouse would be built on a 500-square-foot platform with bamboo as the primary framing and finishing material, as it is local, strong, and light.

The interior walls divide the house into four areas: relaxation room, restroom, kitchen, and bedroom. There are no doors. To illuminate the interior, the roof is constructed of translucent polycarbonate panels. Two of the five exterior sides are screened by bamboo supports while the other three are solid.

The base platform would be constructed of structural laminated wood beams; these would enable the structural load to be transferred to the platform's edge. From each of the pentagon's vertices, an adjustable connection reaches to each palm tree. At that point, special steel friction-style clamps grasp the palm tree.

The profile of the roof is minimal, to reduce structural stresses in strong gale winds.

A centralized hoist is used for water supply and waste; the waste tank remains at a lower elevation, just below the floor.

This treehouse simulates the earth by its exceptional stability.

GREENBERG: It just seems so natural in its conception and integration into such a perfect environment. From a distance it looks like a huge coconut — no one else in the contest has quite so literally imitated nature.

The idea of this treehouse is to bring coconuts back into the palm trees. This treehouse would be perfect for two people, especially honeymooners. It is constructed of natural materials such as bamboo and palm leaves, and the roof is movable. It can be adapted to accommodate groups or families.

The skeleton is a grid shell made of bamboo rods and the roof is covered with palm leaves; wooden rods could be used instead of bamboo. A coating of lime slurry (calcium hydroxide) makes bamboo rods more resistant. The diameter of the bamboo rods should be about 2.5 to 3.5 cm and they should be spaced at about 0.50 cm intervals. The main support level is hardwood, and the floor surface is woven bamboo boards. The top of the shell can be moved with ropes for different weather conditions.

Julie Rouault and Eduardo Gill
46 Spain

SPIRHOUSE

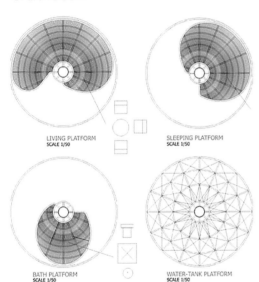

LIVING PLATFORM
SCALE 1/50

SLEEPING PLATFORM
SCALE 1/50

BATH PLATFORM
SCALE 1/50

WATER-TANK PLATFORM
SCALE 1/50

CONCEPT

A Spring . . . Movement . . . A Nest . . . Growth . . . Addition . . . Pack and Go

The house is imagined as a simplified clone of the palm: The trunk is a reinforcement that uses the energy and vitality of the tree in a natural rhythm of growth.

The façade of the house is independent of the platforms where one lives. Its spiral structure allows it to lengthen or to narrow to the rhythm of the tree or of its occupants. The skin is a great wrapping, a filter that protects and unifies, made of a ventilating synthetic or natural weaving. Each platform is an atmosphere, a cell of the house. A stairway is superimposed on the trunk, becoming part of the skin or the spiral.

A natural fence of bamboos planted around the palm determines an external space of transition; the kitchen is located in this public space.

Rainwater is recovered on the plastic roof for the shower. The bathroom is integrated into the bamboo fence and can be moved along the fence as needed.

THE LADY IN THE COCONUT PALM TREE

One day she found herself at the top of a coconut palm tree surrounded by some bamboo plants. She felt happy and fell asleep. The rain woke her up and she had the feeling that the palm had grown. She fell asleep again, feeling good. The noise of a bird building its nest awakened her . . .

ELEVATIONS

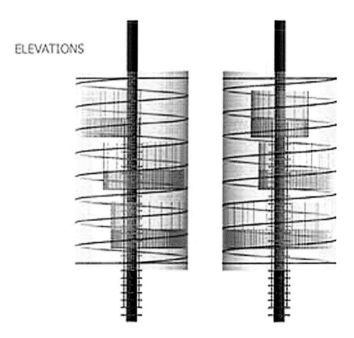

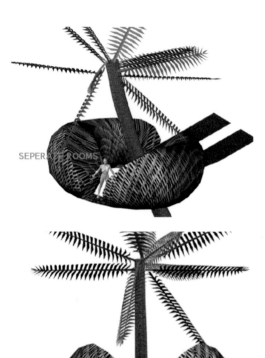

SEPERATE ROOMS

4 DYNAMIC DUO

ALTERED-STATE ROOMS

This design allows play and movement between treehouses when one wants to swing between them like Tarzan and Jane! Rather than designing a single treehouse, the desire was to explore the relationship between a variety of treehouses that would interact as a complex, with activities and entertainment. All groups can congregate at the Welcome Lodge for meals, or have meals delivered to treehouses. Different landscapes are on view from an architecture that changes in relation to the terrain.

Upon arrival, one alights onto a network of suspension bridges, and passes through the Welcome Lodge, which is open-plan and open-sided, resembling a boat deck. Leaving these walkways, one enters the shaded coconut groves and the collection of multi-story Treetop Towers. The four-leveled accommodation towers, shaped like cigars, are wrapped with split bamboo and woven rattan. Each cigar-shaped treehouse is suspended between eight of the tallest palm trees. The lowest level is the bathing area, hovering over the jungle ground; above this is a reception room; on the third level a master bedroom commands a view over the jungle; and the summit is an observatory for stargazing and dreaming. Surrounding the Treetop Towers are the Changing Rooms, bamboo-clad spheres built around a single coconut tree, allowing the occupant to spiral the rooms up the trunk of the tree. Altitude and rotation are gained by a bicycle mechanism. The Dynamic Duo treehouse is cantilevered out over a cliff face and contains two separate rooms that are finished in rattan covered in hard lacquer. This design uses natural materials and native techniques, as it relies on the local community to build and maintain it in harmony with the surroundings.

Ursula Stiglitz

Add to Grow: A treehouse concept.

This design for an adaptable treehouse makes it possible to build various combinations with one elementary structure.

The main box is the starting point, hanging between four palms arranged in a rectangle, joined by bearing timber. Steel connection rings are mounted with steel sleeves around the trees, and joined to the primary construction beams.

The main box is attached to this wooden structure, its secondary beams lying diagonally to the primary beams. An additional bracing is formed by the timber deck at an angle of 45 degrees to the main box, resulting in a small balcony to which a rope ladder can be attached on one side, with a landing for wooden stairs on the opposite side. Inside the main box is only one room for the living, kitchen, and dining areas.

The main box can be enlarged by one to four additional elements: red and blue boxes could intersect with the main box on different levels and would be supported by additional palm trees. A multi-story living, sleeping, and dining area with sliding transparent acrylic panels around the sleeping rooms and bamboo exterior shades engenders intimate rooms that can be separated from the main room by sliding doors.

This treehouse is designed for places that have both dry and rainy seasons; it can be stored when not in use. The roof and floor are designed as two separate parts. The floor is set on the ground and the roof is hung on a palm tree. The walls are flexible. The design allows adjustments for weather in the height and angle of the roof. Interior design is flexible, and the structure can be used in many different ways.

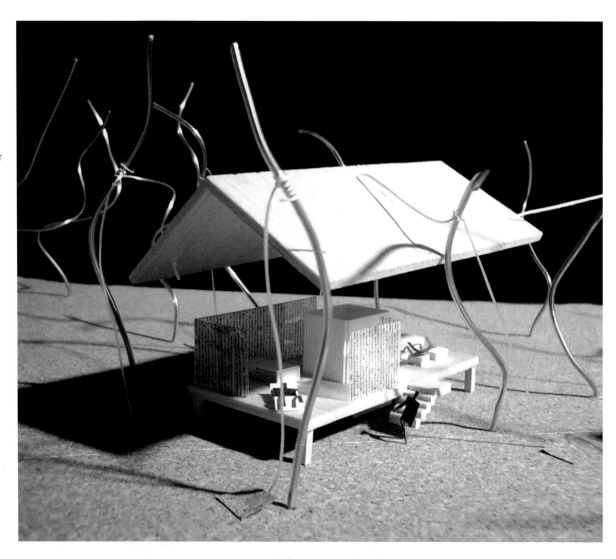

At first we were quite sceptical: Should one open up one of the last beautiful and natural retreats on earth for tourism? Should one confront the proud trees with the "crown of creation"—the human being to spend the holidays amongst them? We could not resist the challenge. After all, there—where we once only carried the tourism in our rucksacks—nowadays the big holiday resorts are located. With our journeys, which many of us have made with little money but much courage, we opened up many untouched areas on earth. Shall we tell everybody now: Stay at home?

Some of our ancestors lived on and around the trees, others found shelter in caves. So we invite the holiday-makers to come back to their origins. Hopefully, a new consciousness for nature can grow from that.

Our outline is influenced by our native country where we have big, wonderful deciduous trees with strong trunks and splendid branches. Our design is based on a construction that avoids damaging the valuable tree: Around the trunk, ring-shaped vertical supporting beams are installed, which are fastened to the main trunk with a padding. At these supporting beams, horizontal planks for the first level are fixed. Stability is guaranteed by crossbeams that connect both elements. Diagonal tie-rods give additional support and build the carrying construction for front awnings made of colorful textile and gauze. In the center of the dwelling level there is a small sanitation facility with outside walls made of polycarbonate plates. The ground-plan is totally open and can be divided up by textile coverings.

Our construction is also feasible without a tree. A post in the ground—an artifical "tree to live on"—is sufficient.

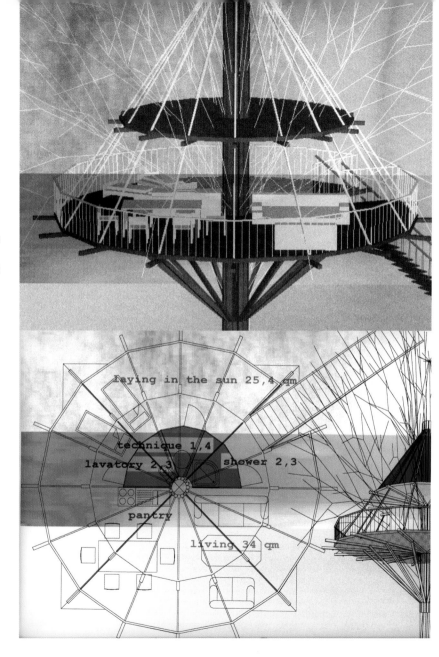

The aim was to create a building that touched the environment lightly and respectfully, using the lessons of landscape and sea to become one with them, while still providing shelter. The tradition of permeable walls in tropical buildings informed the skeleton structure of woven bamboo screens and fabric membrane, allowing the natural elements to filter in, and of shutting tight like a tent.

The project is predominantly inspired by naval architecture—rib and shell structure, mooring, gangways, naval layout, and sails. Stiffening ribs used in boat construction for lightness and strength are used here. In this case, the structural ribs extend beyond the hull and wrap around the top to provide a sense of enclosure while providing an armature for the sliding fabric membrane, lights, solar panels, etc. There is a perimeter beam at the main deck and the lower deck, as well as a ridge beam along the bottom, tying all the ribs together. The treehouse is literally moored to the palm trees around it by a flexible connection of steel cables that prevent the building from moving too much. A nautical bridge acts as the transition between the stair and the floating house.

An abstract interpretation of sailboats is found in the fabric membrane that lines the interior, made of the same material as a lightweight sail; the fabric billows in the breeze and captures shadows, while offering unlimited arrangements of enclosure.

Another metaphor is that of the coconut. This shape plays nicely with the naval forms. The rough, hard, brown exterior contrasts with the softer, fragile, white interior.

A goal of the project was to rely heavily on local materials and labor and to add a few key precision-manufactured components as a contrast. The wood components are in keeping with the local tradition, while stainless-steel bolts and plates add a level of refinement.

My work strives for a balance between the convenience of the synthetic realm and the power of the natural world.

FUNG: This is a project concerned with sustainability, and it is self-sufficient in terms of collecting the energy and resources of the sun, land, and water. It takes on a very serious approach but is also very playful.

GREENBERG: One of my personal favorites as it involves the best of my two favorite worlds—treehouses and ships—with the openness of the deck above and the coziness of the space below.

Wolfgang Leitgeb
and Brigitte Quantschnig

52 Austria

THE PARASITE BOX

The house is a basic unit that hangs on trees. A terrace made of plaited bamboo and net extends around the basic unit. The floor and walls can be customized in various ways. This design allows a strong connection between humans and nature. The house can be easily adjusted for various uses and different surroundings.

It is conceived as a mutable system rather than a unique composition. The in-between spaces of the parasite-box system could be expanded; boxes could be added.

Utilities are located at jungle level and are connected by pulley.

To reach the Parasite Boxes, occupants walk around them on a higher level, pass in front of them on the same or lower level, step down or climb up to enter them.

Suspension bridges connect one tree with its box and rigid platform to another, to allow movement of the trees.

All the roofs can be sundecks, terraces, patios, or observatories.

Parasite Boxes can be entirely opened or closed, and privacy can be regulated by movable screens on all sides of the boxes.

Different materials and elaboration of the screens indivi-dualizes each box; these might be bamboo, locally available composite wood products, oriented-strand boards, medium-density fiberboard, laminated wood boarding, etc.

Parasite Boxes can be occupied by different constellations—couples, friends, families, or groups.

53 Switzerland

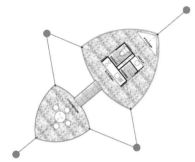

Imagine . . . you find a wonderful place in perfect nature—white sand, coconut palms, lots of birds and flowers, the sea as clear as glass with an untouched reef and millions of fish in different shapes and colors. You find this place and want to stay there!

Bamboo is a native plant in all locations of the competition. The two most important characteristics of this plant are its static quality and its resistance to fire.

You need a special house that protects you, but you don't want to demolish the area with your construction. A favorite example of biological evolution is the shell, combining maximum functionality with efficient use of material.

You want an aviary or a cocoon, hanging between huge coconut palms on different levels. Some cocoons are connected by a suspension bridge, and some by a rope ladder.

The mention of a treehouse evokes contradictory images—of monkeys climbing from branch to branch, of children in their playhouse, of a hermit's spiritual abode. But the essence of a tree is the evolution of life. This is the element we have explored by splitting the treehouse into different levels and structuring it around a staircase. The staircase anchors the levels and symbolizes the vertical ascent of life's evolution; it is the core of the design, which spirals. The open decks are shaded by pergolas or by wing-shaped waterproof membranes that make the whole structure dynamic, as if it might fly away at any moment. A single tree, wherever possible, has been used for support, but spaces that are supported on more than one tree have been designed on the fixed- and roller-joint principle.

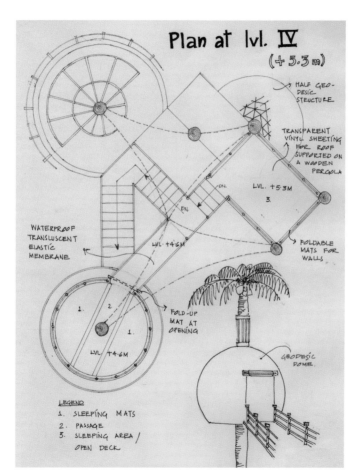

Plan at lvl. IV
(+ 5.3 m)

HALF GEO-
DESIC
STRUCTURE

TRANSPARENT
VINYL SHEETING
FOR ROOF
SUPPORTED ON
A WOODEN
PERGOLA

ON.
DN.
LVL. +5.3M
3.

WATERPROOF
TRANSLUSCENT
ELASTIC
MEMBRANE

LVL. +4.6M

FOLDABLE
MATS FOR
WALLS

1.   2.
1.

LVL. +4.6M

FOLD-UP
MAT AT
OPENING

GEODESIC
DOME.

LEGEND
1. SLEEPING MATS
2. PASSAGE
5. SLEEPING AREA /
OPEN DECK

What is it about a treehouse? The feelings a kid gets in his backyard hideaway—solitude, seclusion, hiding, escaping, and a kind of voyeurism—nobody can see you, but you can see everything.

Coconut palms provide the structural elements. The first level, a few feet above, is mainly an open platform for utilitarian functions. The second level is a vertical layering of operable membranes that allows the boundary between interior and exterior to disappear, to be blurred, or to be opaque. The outer layer is a simple retractable insect screen that surrounds nearly the entire structure, creating a comfortable exterior communal space that rings the more private sleeping area. This space is always open to the surrounding treetops. Operable membranes comprised of wood-louvered sliding doors can completely disappear and render these inner spaces "wall-less," or be closed for total privacy. At the center of the second level is the most private of the spaces—the compost toilet and shower; this space is enclosed on all sides by an opaque membrane.

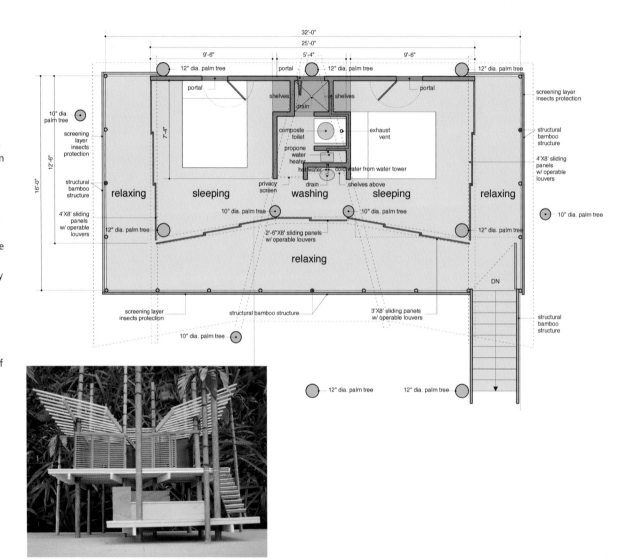

Kenneth McCown and Kevin Hinders

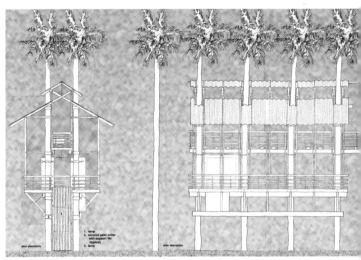

THE PRIMITIVE TREE-HUT

What does an architect do? The ancient Vitruvius said that architecture is an art form that blends function, structure, and delight. The architect can bring the client's fantasies to light and fruition. In this case, we are the clients creating our own fantasies.

The process began by oscillating between the logistics for a structure that could not touch the ground and the notion of creating a primitive getaway. Could trees be located and replanted? How would the swaying of a palm tree change things? The tree must have a gangplank to connect earth and sky for those who can no longer climb a rope.

A "kit of parts" could be easily understood by local builders, allowing joinery to be determined by the local cultures. As architects, we would then become facilitators of collaboration, making unfinished systems that can be expressed locally.

The trees are pinned at the bottom of the structure but remain free at the top so that the crow's nest doesn't move with the tree. The Hut remains a simple frame that doesn't distract from the beauty of its surroundings.

Jurgen Schimmelpfeng, Claudia Wolsfeld, Sandor A. Konnyu, and Stefan Kamphans
57 Germany

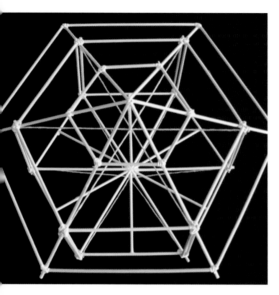

Our treehouse hangs in a group of about twenty coconut palms at a height of seventeen feet. The rooms are arranged around an open terrace. Every room has its own function and hangs flexibly between the trees. Leaves protect the house from rain and sun. The five- and six-cornered rooms fit harmoniously in nature; they are constructed of regional materials such as bamboo and rope. Ropes between palms carry each room by its top corners. Ropes are connected to stainless-steel sleeves at the palms. The ropes under the rooms are stabilizers.

Crina Sebastiana Ion
58 Romania

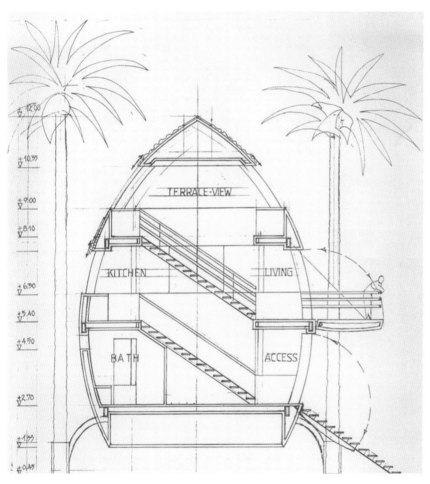

TREEHOUSE IN HAWAII

Coconut flower opening toward the view, over the beach and over the sea, floating above the earth, closer to the green leaves of the coconut tree, closer to the birds, closer to the sky . . .

This could be a coconut flower community, connecting all the treehouses by suspended passages. It is an easily dismantled and removable structure, with a circular platform suspended on four points on four coconut trees, through four steel pipes to four steel sleeves, supporting the other two floors; and a cage of ribs made of: 1) laminated wood with aluminum or steel joints; 2) metallic profiles; 3) other local light and resistant materials that can be fashioned into curved ribs. Finishing materials are bamboo, sugar canes, raffia for ties, etc. Furniture is made of wicker, bamboo, plywood, raffia ties; bath and kitchen equipment are of metal or plastic. Rainwater is collected and recycled, solar energy heats water, and an underground plumbing system is installed with pumps and pipes.

If people were like trees they would have their feet planted firmly in the ground and carry their heads in the sky, in heaven.

The treehouse is connected to the trunk with two platforms. The lower platform is for sleeping and sanitary areas and the upper platform is for living and cooking.

The roof is a flexible, tentlike structure based on the classical wood constructions of East Asia and Oceania; it is built of eight wooden poles that are bound in pairs to form the lower platform. The poles form an enclosure at the lower level and extend over the upper level into the palm-leafed roof. Depending on the angle, the viewer sees the roof as a sail or as a crown. Let's be carried away by the palm-sail, or crowned by the palm roof!

Accessed by an external timber staircase, the house is suspended five-and-a-half feet above the terrain on six coconut trees by means of stainless-steel cleats detailed between the coconut trees and the nodes of a hexagonal structural base frame. The structure consists of a ground floor and an upper floor. The external staircase leads to an entrance porch on the suspended ground floor, which connects the interior. A double-flight internal staircase, set between the kitchenette and the washroom, allows an open link to the upper floor, where a private compartment doubles as a living and sleeping space. Furniture is tucked into side extrusions, making the stairwell the focal object of the space. The compartment opens out by means of sliding panels to the open terrace. Overhead cabinets on the upper floor provide sufficient storage room.

The design focuses on good cross ventilation through adequate openings in the walls; these openings are also the source of daylight. The hipped roof explores a cover of combined vinyl and rubberoid sheets.

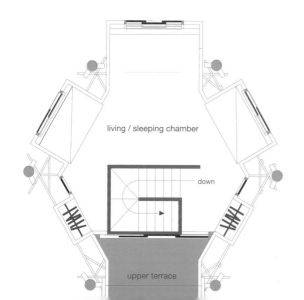

living / sleeping chamber

down

upper terrace

Kendel Architekten
Jan Hugo Kendel and Philipp Kendel
61 Germany
Top Ten Finalist

This is a treehouse that can be simply assembled and produced even in areas with very low-level infrastructure. Alternatively, it can be prefabricated as a high-end pack-and-go version for export into off-road regions.

This structure allows a controlled movement that keeps a good distribution of forces on the structural parts; this is a crucial aspect when designing for supports that are moving and growing.

The materials for the "stage of high life" treehouse are two palm trees, three different kinds of metal joint modules, and a few meters of stainless-steel cable. For the "cocoon," any material such as palm leaves or coco fiber mats can be fixed on a simple lattice-grid shell of local timbers. For the high-tech pack-and-go version, glass fiber rods would be used instead of wood and the cocoon cover would be made of modern parachute materials and Gore-Tex tissues.

The treehouse combines the principles of a spinal column and a hammock, which means it will swing very lightly and twist with the wind and the trees, but will remain stable at anchor in the air. The overall size of the treehouse can vary according to the distance between trees and their trunk

dimensions, from a single/double room to a family apartment. To create a "lonesome nest," two trees are needed; to create "community clusters," several houses are fixed to one central tree and an according number of surrounding ones.

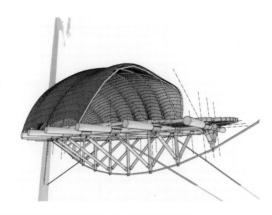

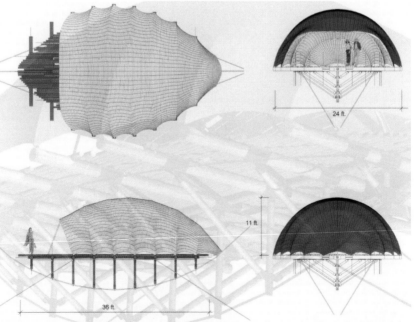

24 ft.

11 ft.

36 ft.

BURTON: Here is a balancing act between temporality and permanence, though nothing is really permanent. You have a totally different view of the universe when you are up there in the tops of the trees and swaying with the breeze. When you try to make something solid and permanent you are going back to the thought of a regular house and those patterns.

FUNG: It looks like a hammock, swinging in the trees, which fits so well with the whole idea of being in a tropical island, giving you the notion of being on vacation. One of the more contextual projects.

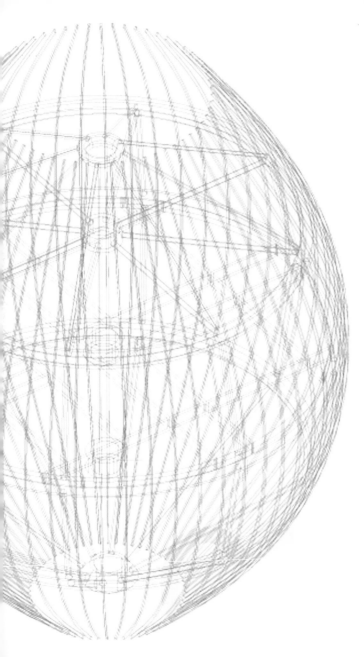

Cristian Stelzner

62 Spain

THE FAT PALM TREE
My intention is not exactly to
build a house on a tree, but to
transform the tree (a palm in this
case) in a way that produces what
we are looking for—a space to
live in. I began to work with the
image of a silk cocoon until finally
I arrived at "the split trunk." The
trunk is cut in thin strips, opened,
and made to form the sphere that
can be used as a house or . . .

GREENBERG: The
geometry of the skin
seems like a work of art.

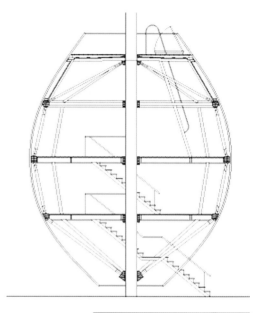

The three levels of this house emerge from one another so that the whole volume of the house can be felt from any level. The first level is a deck to be used as living/dining room. The second and third levels are bedrooms with adjoining decks, oriented toward the sea. Local materials, mostly bamboo, are used for construction, screening, and finishing. The sloped roof is made of colored corrugated tin.

Nest—a secluded shelter or hiding place; a snug place.

The idea is to create a hideaway for a couple. The spatial arrangement is a split-level scheme consisting of insular sleeping areas and open living areas. The construction consists of dimensional lumber, fastened with metal brackets and hardware. Translucent panels provide shelter from rain and wind.

001 access ladder
002 stainless steel bracket
003 existing tropical tree
004 wood railings
005 translucent panel

kitchen area

living area

sleeping area

bathroom area

scale 1/4"=1'-0"

Mikhaïl Borchtchevski and Elena Kondratieva
65 Russia

Our treehouse is fixed on an axis of four coconut palms bunched together, each 45 feet in height. Four wooden rings, placed at vertical intervals, encircle the bundle of palm trees and are connected by flexible cords to the horizontal members of the treehouse. This creates a system of rotation originally invented by Leonardo da Vinci. This arrangement allows the house to rise and fall, according to the whim of the inhabitants, and introduces pleasant variety to days spent in the treehouse.

Solar batteries can be built into the plastic roof. The floor area of 781.9 square feet includes gallery, kitchen, drawing room, two bedrooms with bath, and ecological toilet and shower.

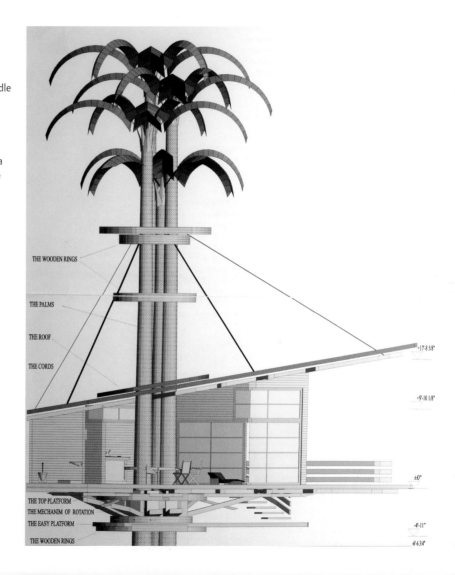

THE WOODEN RINGS

THE PALMS

THE ROOF

THE CORDS

+17-8 5/8"

+9-10 1/8"

±0"

-4'-11"

-6'-6 3/4"

THE TOP PLATFORM
THE MECHANIM OF ROTATION
THE EASY PLATFORM

THE WOODEN RINGS

Olga Sidorina, Arkadiy Tsaturyants, and Aleko Bahkatov
66 Russia

Soaring above the ground, these gigantic coconuts serve as dwelling capsules.

VERSION 1

A seven-meter, three-level module hung between palms on a triangular platform floor is firmly fastened to the trees with steel couplings. An open square rotunda on wooden poles accesses lower levels across suspension footbridges. This dwelling is suitable for one or two people.

VERSION 2

A block of three coco-capsules forms a three-room apartment for three or four people. Each room is connected with the others by an inside court. Construction of this house is similar to the first version.

The walls of coconut houses are wattled rattan baskets. A wooden framework of vertical bamboo and horizontal plywood is interlaced with coco rope or other suitable natural material. The inside wall finishing is white clay plaster. The open terrace, balconies, and inside courts are fenced with tied bamboo and clay plaster. Trimmed straw made of coco leaves is also employed. Tasseled roof sails, like a tent, protect from rain and sun.

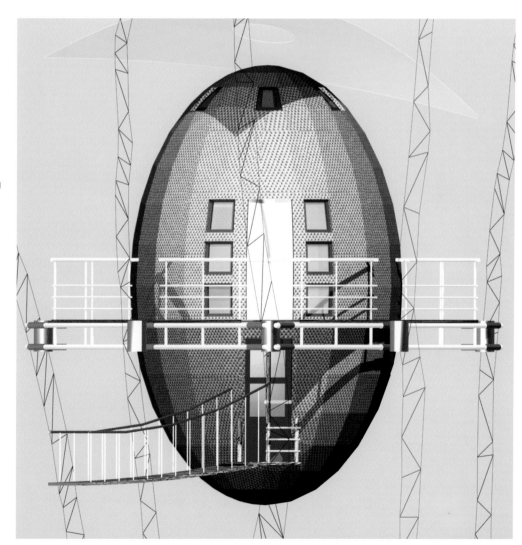

Marianne Bär, Beter Dorr, and Jens Kolb

67 Germany

Top Ten Finalist, Third Place

Unceasing wind and coconut palm trees characterize the tropical islands on which the treehouses are to be built.

A treehouse using coconut trees as the primary bearing structure cannot be static; it must be able to deal with the tree's movement. Thus, we have designed objects to be hung between at least three palm trees each, and suspended by a secondary bearing structure such as bungee ropes. A system of inflating the treehouse like a balloon was also developed.

The natural shape of a coconut is the starting point for a transformation. This transformation is determined by the three necessary palm trees, which are considered deforming forces. The use of high-tech materials such as a framework of aluminum or carbon fiber contrasts with the natural, original shape.

To maximize the impression of objects hovering in the air, contact with the ground is reduced to a minimum. Access to the treehouse is by pulley block.

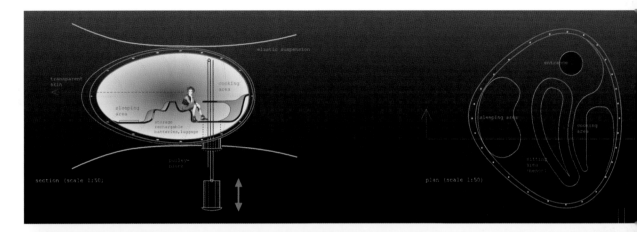

BURTON: I think that this project, with its spherical egg, is like some strong web from a spider—the movement and flexibility and getting up by a pully system is so unique.

FUNG: Coconut in a treehouse, out of a bottle. You inflate. This is genius—that it doesn't exist and then it does. Also using low-tech, but the look is very high-tech. Stunning.

GREENBERG: This particular entry was the finalist most often published in magazines worldwide and is probably the best example of a treehouse for the new millennium. Though not all the details have yet been invented to realize this dream, the architects assure me that they are working on it.

gentle movement

Keith Moskow
68 USA

This treehouse floats lightly amongst the trees.

The bath and a compact kitchen are placed to the north. To the east and west are sleeping areas. To the south is an open area housing a large table for working, reading, and eating. The four components ring the central platform, creating a protected outdoor room. Tikki torches in the four corners are lit at night. Sparks twinkle into the tropical air.

Each of the four components is 8 x 16 feet for a total interior space of 512 square feet. The platform provides an additional 256 square feet of outdoor space. The treehouse is constructed of native materials—bamboo, local hardwood, and woven palm fronds.

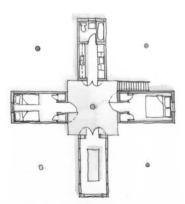

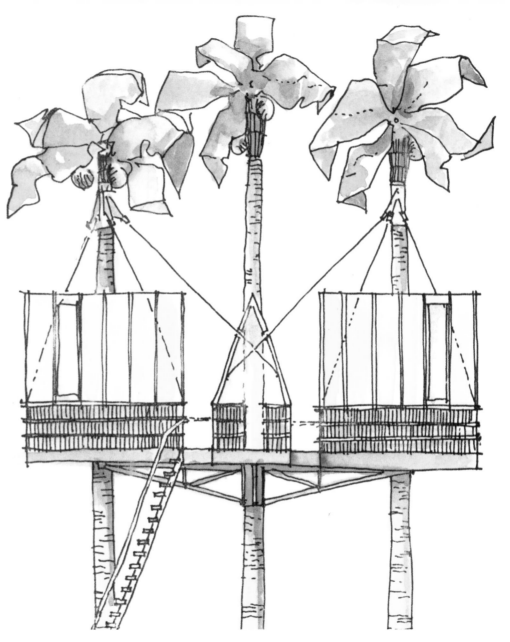

Spurning indigenous materials such as bamboo and palm-frond roof for a more contemporary use of materials. The basic structural connection to the palm trees, however, remains the same and is probably a unique and satisfactory method. First, attaching low on the palm tree and then fanning out the connections to the beams.

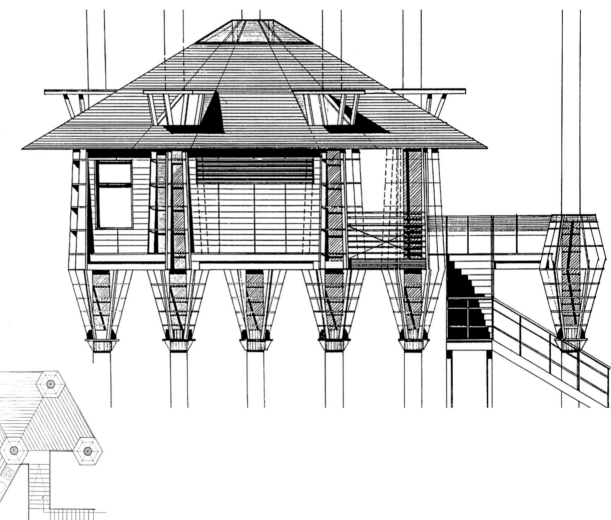

Masakra

Tomasz Suchy, Sonia Pardi, and Anna Pliszkowska
70 Poland

We designed a little house on a tree, meant for people who live actively and enjoy adventures. We were inspired by natural forms like coconuts and dewdrops.

The structure allows the opening of the whole house to the view. The design features construction-steel supports and steel hoops, waterproof covering material, and an openwork bamboo layer. The tree is reinforced by ropes anchored to the ground. We provide a shared sanitary and kitchen complex on the ground among the treehouses.

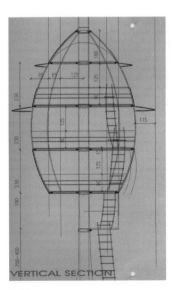

VERTICAL SECTION

THE SEVEN STEPS OF TREEHOUSING

Step 1: I like to be above the ground, under the sky, between the leaves, feeling the wind and the movement of the trees. I want to seek a little quiet to collect my thoughts and gain a little wisdom. I'm not an animal or a saint, living on top of a column. I am a human being.

Step 2: I don't want to climb, I like to walk nicely and easily, to move slowly and safely, to build a bridge between earth bound and sky taken, feeling that I can choose, and that no one pushes me to be at the one or the other place.

Step 3: And when I arrive I will at one point have a wonderful view over the sea, and when I turn my head I can look at the landscape from a little above and many things become more and more clear.

Step 4: I can get out my books, sitting here above the ground in the shade, reading on the big balcony. Or I rest on the bed, having a dream.

Step 5: When I wake up and feel hungry, I can walk a few steps down to the kitchen and prepare a meal; while cooking, I turn around and go to the small lower balcony, looking for friends or guests arriving or playing in the garden.

Step 6: In the garden where the flowers attract butterflies, I can take a little private shower, hidden behind the bamboo fence. I don't need much, I only need the right place to be.

Step 7: I like the house looking like an animal, resting and gazing over the sea. I don't want a square form, I like movement and curves. And . . . have fun . . . for yourself at the top of a tree, hide and peer out over the landscape.

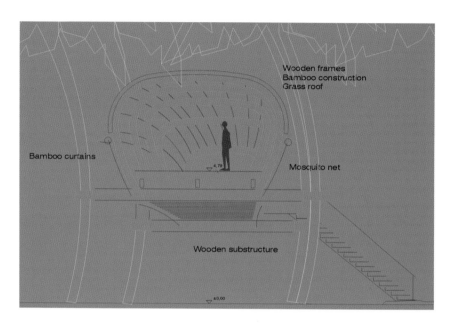

Wooden frames
Bamboo construction
Grass roof

Bamboo curtains

Mosquito net

Wooden substructure

Xisca Monjo Ballester and Jordi Rodriguez Arroyo
72 Spain

We want to have our own tree growing from the earth to the sky so we can reach to the top. We want to plant our tree in the middle of a coconut palm forest. Our tree grows as a shell, getting closer to the other trees as it grows.

In the middle of a coconut landscape our coconut is hanging among coconuts.

Our coconut is our nest, the place where we feel safe, a suit made to measure, wrapping us as we sleep. We feel the movement of this flexible structure.

Shell and nest are two different structures: The shell is positioned on the ground, without holding onto the tree trunk; this allows it to be stable when the tree sways; the flexibility of the materials ensures this. The shell is the space where we sit, move, reach. The nest is a smaller space made of netting, which can be covered with palm leaves.

Here we have the minimum structure from which we can begin to understand the surroundings and construct our way of living in the place.

In the end, our tree becomes our home. We are part of the landscape.

Tonight is really calm.

Felices sueños!

Ground Floor Plan

This treehouse surrounds central trees in four separate units, together making one treehouse. There's a large entrance room, a unit for sleeping, and a dining unit, which are connected by stairs. Four to six treehouses dotted along the seashore make a nice village of treehouses.

Just plain poetry in repeating
sensual curving forms. A skeletal
structure reminiscent of both
natural and modern forms.

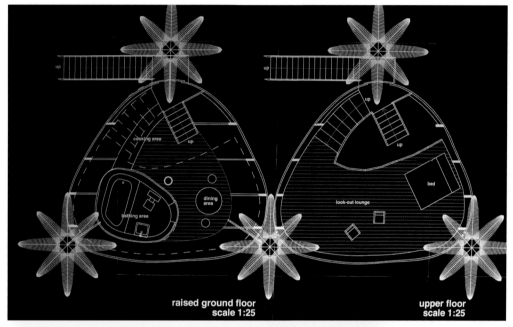

raised ground floor
scale 1:25

upper floor
scale 1:25

75 Daniel Krautschat and Angelique Zierenberg

The treehouse competition was part of the first semester in the design department of Professor Bernd Huckriede at the Brandenburgische Universität Cottbus. The design process was directed and assisted by the architects Minka Kersten and Axel Büther during winter term 2000.

The approach to architectural space began with exercises detached from the common way of thinking about architecture. These projects were the first steps and involvement into the perception of space, scale, dimensions, and materials.

Step 1—*Cover for a Plant*: Different plants were analyzed and a particular cover was designed, referring to the character, the needs, and physical aspects of each plant.

Step 2—*Capsule*: The structures and different textures of fruits were researched in various tests, in sections, imprints, and minute views of different scales by microscope. The results of the analyses were transferred into the idea for a small capsule that provides enough space for a group of three to four persons. A series of creative projects were presented.

Step 3—*Treehouse as a Small Housing Project*: The sequence of exercises referring to plants is continued in the treehouse competition. A characteristic small house for two or three persons was realized with bamboo material, constructed with simple bounded connections. The design process was based on working models in the scale of one to ten.

Five of these projects, shown here and on the following pages, were submitted for the treehouse competition. These projects show the variety of topics and ideas the students were involved in: ideas of space in the treetops, the flexibility of the house to create an agreeable space, the qualities of the material of bamboo with its structure and perforation, which allows the play of light and shadow.

The design by Maren Bückmann and Arne Eicker transformed their interpretation of a poem into a private treehouse. A space, created by a pattern of linen sails, gave the project a light and changeable appearance. Observing the surrounding landscape became a theme for the house: Like an eyelid, one part of the treehouse can be opened toward the landcape.

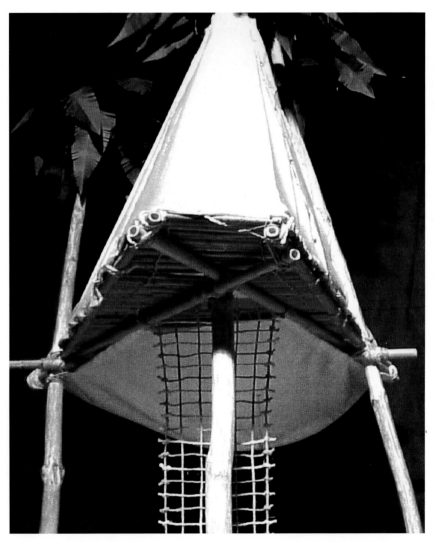

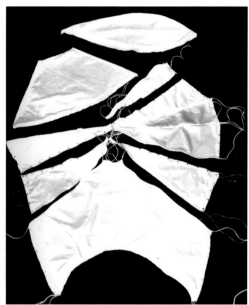

76 Maren Bückmann and Arne Eſcker

77 Sabine Zweifel and Anne Wrobleski

78 Ronny Kletke and Doreen Paprosch

79 Lars Offergeld and Sara Sohnel

Treehouse: living in a tree—the breezes—the moist air—the swaying of the trunks—the dappled light—the smells—the rain—the sounds—the color—the sea—the land—the natural and the man-made—surrounded by hammocks—deep shade—receding into nature—lightweight—meandering through the trees—the fallen log—the found space—the fun . . .

The treehouse is inspired by the idea of the fallen log. Its long meandering plan acts as a found space for the vacationers, in the same way that a fallen log provides a home for the various animals and insects that are lucky enough to find it first. The plan shown is one possible arrangement for living, however the "found" space is completely flexible to accommodate each group of people, separated only by the curved stone wall that rises above the ground and alludes to the limestone hills and mountains that typify the Vietnam landscape. The stone service core provides a connection between earth, wind, and water, both visually and philosophically.

Overhead is suspended a lightweight waterproof membrane, with varying degrees of protection, some areas with large overhangs, some smaller (much like a tree itself). The spaced-out decking board floor is surrounded entirely by a giant hammock running the full length of both sides of the house.

The emphasis for this organic treehouse is on craftmanship and allows the construction team and the client to design as he/she goes. One version may be 82 feet long. The floor plan may twist, rise, or fall any way in order to suit the existing trees and the landscape.

Visually and physically the house recedes into the environment (the trees), embracing it and celebrating it, allowing the occupant to have maximum contact with his/her immediate environment.

This treehouse is not a set, static model, it is based on an idea of how, (maybe) to live amongst the trees.

We want to rest peacefully in the shade of the wild to keep our soul strong and free. We don't need much space or comfort; we want to feel the heat and cold as the tree does, and the serenity we once felt in the womb.

The structure most independent of environmental stresses is the strongest, lightest, most flexible one. Flexible technical screws and lightweight fabrics are employed with an umbrella structure to endure monsoon conditions.

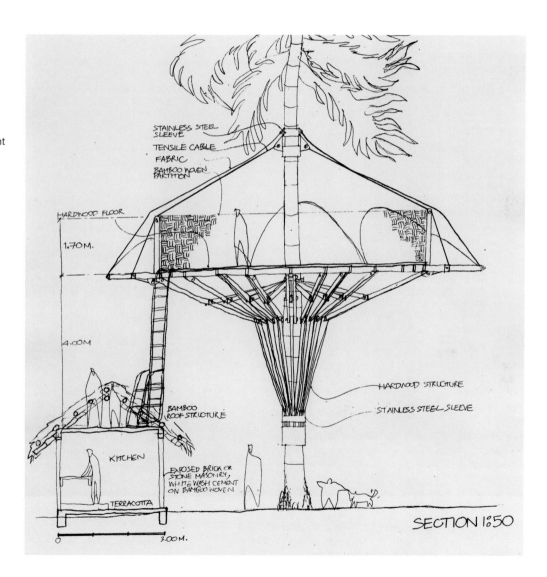

STAINLESS STEEL SLEEVE
TENSILE CABLE
FABRIC
BAMBOO WOVEN PARTITION

HARDWOOD FLOOR

1.70 M.

4.00 M.

HARDWOOD STRUCTURE
STAINLESS STEEL SLEEVE

BAMBOO ROOF STRUCTURE

KITCHEN

EXPOSED BRICK OR STONE MASONRY, WHITE WASH CEMENT ON BAMBOO WOVEN

TERRACOTTA

3.00 M.

SECTION 1:50

LE COCON DANS LES ARBRES

Reading *The Baron of the Trees* by Italo Calvino inspired a treehouse that is modular and progressive. We worked with two types of space: The space of intimacy and relaxation is the modulus cocoon; the space of moving and community life is the modulus platform.

The basic treehouse is one cocoon. The next progression would be the addition of a second cocoon and a bridge. A more comfortable house would combine three cocoons and a unifying platform. One more cocoon and one more platform could be added. And so on. Every modulus is itself modular, allowing changing uses.

The unifying platform is open to nature and landscape, covered with polycarbonate plates and bamboos. Rollup hangings modulate light and protect against weather; these would be handcrafted locally in a variety of colors. The cocoons are each created around just one tree trunk and the platforms are slipped laterally by way of a pin system.

We maintain proximity with nature by opening the living area to the surrounding landscape and by passing the palm trees through the cocoons. We maintain proximity with the people by using local crafts.

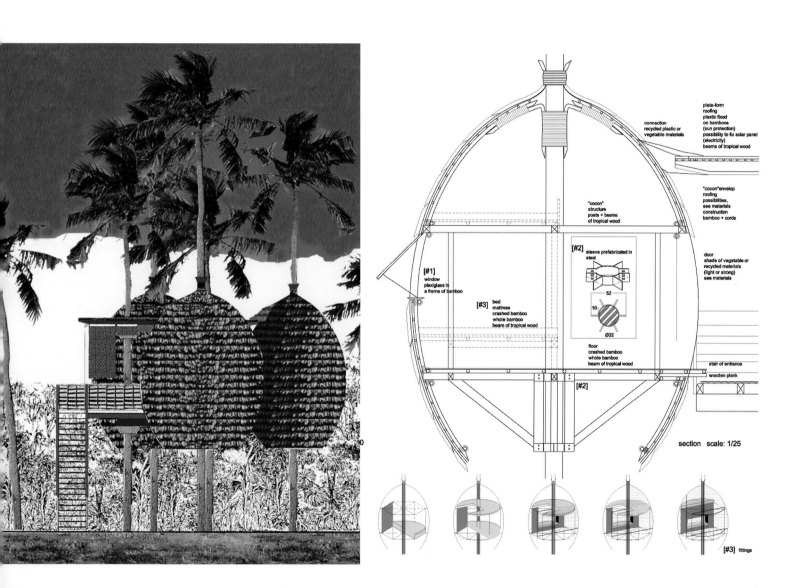

connection
recycled plastic or
vegetable materials

plate-form
roofing
plastic fixed
on bamboos
(sun protection)
possibility to fix solar panel
(electricity)
beams of tropical wood

"cocon"
structure
posts + beams
of tropical wood

"cocon"envelop
roofing
possibilities,
see materials
construction
bamboo + cords

[#2]
sleeve prefabricated in
steel

[#1]
window
plexiglass in
a frame of bamboo

door
shade of vegetable or
recycled materials
(light or strong)
see materials

[#3] bed
mattress
crashed bamboo
whole bamboo
beam of tropical wood

52

10

Ø32

floor
crashed bamboo
whole bamboo
beam of tropical wood

stair of entrance

wooden plank

[#2]

section   scale: 1/25

[#3] fittings

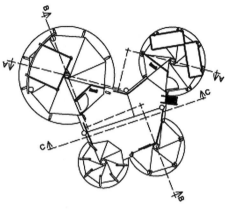

The organic coconut shape illuminates the relationship between the project and the spirit of the site (genius loci). Thus, the "coco-room" hangs on the tree trunks.

The design transforms domestic architecture from static shelter to dynamic habitat. The plan follows the natural disposition of the palms; thus, it isn't a production model, but rather a typology.

A single trunk carries the weight of the house, thereby canceling wind stress. The house is connected to the trunk by steel-edged elements with hooks for horizontal elements. The top two "coco-rooms" are panoramic terraces. The third "nut" is the kitchen. The fourth room, on the ground floor, is equipped with ecological WC and shower: pipes are covered with bamboo.

Alex Hirsig

84 USA

This structure follows the nature of organic growth. The furnishings and objects within should be easily adaptable to sudden shifts in mood or weather. The local artisans will provide traditional crafts to fill these treehouses.

The treehouse is designed as a composition of larger tensile structures and smaller pieces come together to form a waterproof canopy. The twelve structures are of marine canvas and flexible wooden poles creating a lightweight and durable skin.

Caroline Trolle, Vendela Thelander, Caroline Fleming, and Fredrik Ekman

85 Sweden
Top Ten Finalist

Our proposal is easy to build with local materials and adaptable to all conditions. Combined, the cocoons create a cluster. Provided that they are placed with care, they avoid exploitation of the site.

Each building is attached to one tree and connected to another by a suspension bridge. During rain, the whole structure can be covered with plastic and stored on top of the cocoon. The floors are upheld by lumber shaped like an umbrella, attached to the tree trunk and fixed by a steel collar. At the end of each timber, bamboo sticks are attached to shape the cocoon. The amount of daylight can be adjusted by hatched openings. At night, the cocoon forms glow from inside, creating a ribbon of lanterns on the horizon.

The cocoon formed on a grass stem by the little spittle bug was the source of inspiration for our construction.

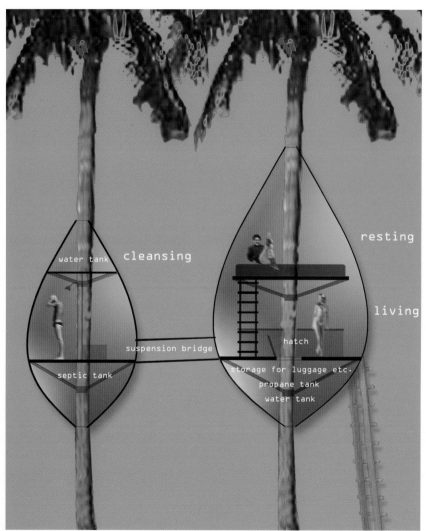

cleansing

water tank

suspension bridge

septic tank

resting

living

hatch

storage for luggage etc.
propane tank
water tank

BURTON: When you think about the form of these you think about *On Growth and Form* by d'Arcy which takes a look at the geometry of all forms in nature and biology and plant life… what a better idea than to camouflage it somehow in the forest such as the various materials they plan to build out of. Even the cloth could be like a giant insect. If you look at the section it is like a tent … I love that this is very simple and very elegant.

GREENBERG: It is very Scandinavian as the textiles become part of the architecture. After meeting with the three female architects I was so pleased to understand how this was such an interesting study in group creativity. One free idea leading to another in another mind and then worked on collaboratively. I suspect that many of the projects with multiple architects were also done in this way, which is not a particularly normal practice.

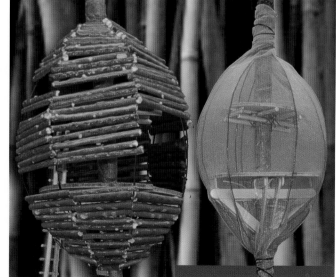

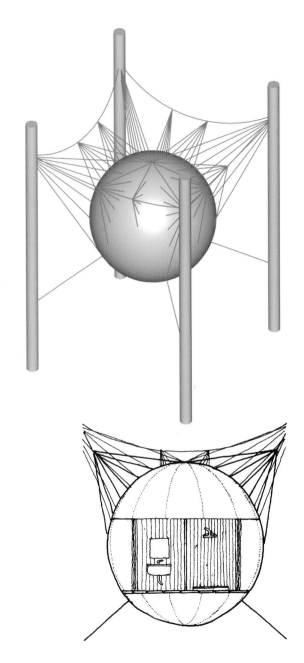

Four palm trees are chosen for each inhabitable pod, regardless of size. Natural ropes are stretched between the trees to create a square frame to which suspension cables are attached. Suspension ropes are hung from the main ropes and knotted to the framing ropes on the pods. The precise location of the knots makes a complete circle in the vertical and horizontal planes.

A membrane is stretched between the framing ropes to make the enclosure of each pod. The soft or hard floors are suspended inside from the framing ropes to complete the structural system. The entire system is made of only three materials: ropes, fabric or membrane, and suitable materials for the flooring. A vast hammock is stretched between the inhabited pods for easy access and movement between the pods. The size and shape of the stretched hammock are customized according to needs.

By living under unusual and even harsh living conditions, one may discover the meaning of life. One can crawl and bounce around on the hammock, or sleep in the spherical pod looking at the stars. An adventurous spirit would infinitely expand the complex, to create a kingdom in the treetops for people who have not forgotten to dream.

THE HAMMOCK HOUSE

Hovering like a colored cubic fog between its palm columns, the Hammock House is suspended between four trees by a steel cable tensile structure. It is a small, simple structure for two people's basic needs for life in the trees.

The cubic form is enveloped by a thin veil of colored mosquito screen. The color of the screens changes from house to house, as in "Hi! I'm the blue house, which one are you?" The interior space is formed by the pinwheeling of horizontal and vertical planes, some of opaque hardwood board, some of translucent canvas, and some of filtering bamboo slats. Interior canvas partitions are operable on sliding tracks, as is mosquito screen on the ocean side.

Tree dwellers enter the hammock house by way of rope ladder and a simple hole in the floor. The first level is a hangout, the second level is a sleeping platform, and the plumbing is located on the ground level for communal use.

Solar panels are integrated into the bathroom-house roof, and composting toilets are used.

Made of traditional indigenous
materials alone, the shapes,
forms, and colors provide a simple
solution. It's sort of a tent with a
bamboo structure. The skin would
be of local artists' colors and
designs. The simple form allows
for protection at both ends.

NOTE: Entry arrived late for the competition,
therefore information is incomplete.

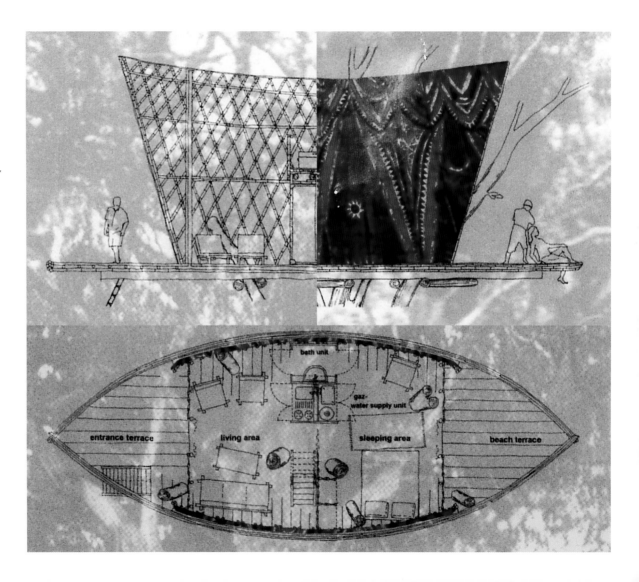

bath unit

gaz-
water supply unit

entrance terrace   living area   sleeping area   beach terrace

Set in a clearing in the jungle brush, one approaches the complex through a break in the trees adjacent to the beach. Attracted by the glow of sunlight, a visitor finds himself at the entrance to a treehouse community.

The circular clearing forming the center of the community is sacred. The sacred space is bounded by a circulating path, gently spiraling upward, providing access to each unit. The path is supported by native ropes and hung from a rope tension ring high in the air, which further defines the sacred space in three-dimensional form. Arranged equidistantly around the ring are five treehouse units. Three are private living units, and two are shared facilities.

Each unit consists of a supportive structure of three coconut palm trees, planted to form an equilateral triangle. Two primary beams of native hardwood support each floor and roof structure. Each pair of beams originates from one of the trees and at its midpoint straddles an adjacent palm tree. The canti-levered portion of the beams contains secondary supports that are covered by a wooden deck. The same system is employed for the roof, with corrugated tin.

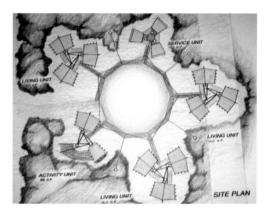

The second unit contains utilities for the community, with partially enclosed showers and composting toilet. Water is heated by propane gas and solar energy.

The Activity Unit is located at the apex of the circulating path, high above other units to provide views. The structure is similar to the living units but is simplified and "exploded," with each deck five feet above the previous one. Each level is dedicated to one of the elements: the lowest is for use during rainstorms; the second is equipped with a firepit covered by metal sheeting and a layer of dried earth; and the highest level is a viewing space 35 feet above ground with a cargo net attached to the beams; one climbs over the beam and into the hammock created by the net.

GREENBERG: I think the model of this treehouse shows it to be great fun, a great design, and certainly unique in its structural integration with the palm trees.

Marco Bruno Periera de Mateus Silva and Vera Gomes Cirilo

90 Spain

This design is based on an analogy between the development of the volume and the growth of a tree. Nature builds in harmony with existing elements, keeping fluidity between what is new and what is already there. This concept led me to the birth of the House of Leaves.

Once the basic principles were defined, the volume grew and took its own shape. The growth occurred in a continuous and fluid relationship between inner and outer spaces, establishing the desired link with the natural environment.

At the entrance to the house on the lower level, there is a space for relaxation, with a cooking and storage area. In the upper level, two sleeping spaces contrast with the space below.

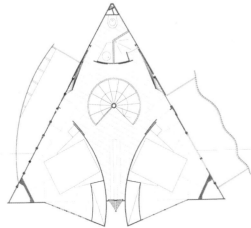

André de França e Campos
91 Portugal

It is necessary to build without destroying nature!

I propose two wooden "boxes," opened at their tops. Both boxes function autonomously—one is for greater privacy (bedrooms and bathrooms), the other functions as the social area (living room, dining room, kitchen).

In the center rises the ladder, the one element of connection between the two boxes.

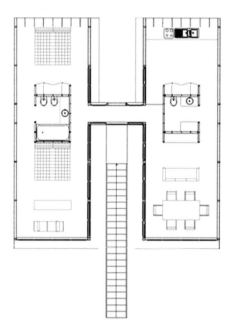

Tim Rieniets

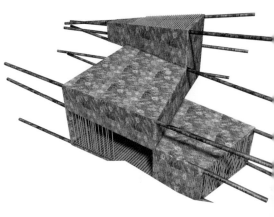

FORM FOLLOWS TREES

A treehut for tourists must be romantic; it must provide the treehouse dream. The random pattern of the tree trunks in the grove is the natural condition for generating numerous individual treehouses. A simple construction of beams spanning tree to tree creates a unique form for each house. Depending on the given pattern of trees, each house will have a different organic form. Each form has a different ground plan and different spatial qualities.

Wall coverings of chicken wire and leaves sandwiched together make the houses appear homogeneous. This covering provides good ventilation while incorporating pipes for facilities.

Montevideo 777

Mauricio Corbalan, Gustavo Dieguez, Lucas Gilardi,
Daniel Goldaracena, Pio Torroja, and Santiago Costa
93 Argentina

AIRPLANTREEHOUSE
FOUR PEOPLE
SIX PALMS

SHIPTREEHOUSE
SIX PEOPLE
TEN PALMS

LOCOMOTREEHOUSE
SIX PEOPLE
ELEVEN PALMS

After the wreckage, where is it safe? Weather Forecast (Province of Hue, Vietnam):
A: We've heard the forecast. The news isn't good. Force seven winds are expected.
B: No problem! We can lower the treehouses, they're like wooden flags.
A: So after the storm we can hoist them again!

Charlie Don't Surf:
A: The corrugated tin sheets were too big to fit into the canoe.
B: All we have is bamboo canes and wooden boxes! The plastic sheets won't arrive until next week . . .
A: The rain has washed away all the detail drawings, but if we can remember the figures and the structural schemes, we can go ahead anyway!

Gentils membres (Kind friends):
She: What a good decision to come to a club in the tropical boonies!
He: Yeah, but . . . the country seems devastated by monsoons . . . Look! Ships and planes have been blown into the trees!
She: No, dummy! Those are the treehouses we were headed for.
He: Well, at least we're out of this world.
Child: I don't wanna go! I don't wanna leave here! I want to take all these toys with me!

No more boring waterbeds (swishhhh! the wind whistles):
She: Ohhh! You rock me sooo gooood!!
He: It's not me! It's the f---ing house!

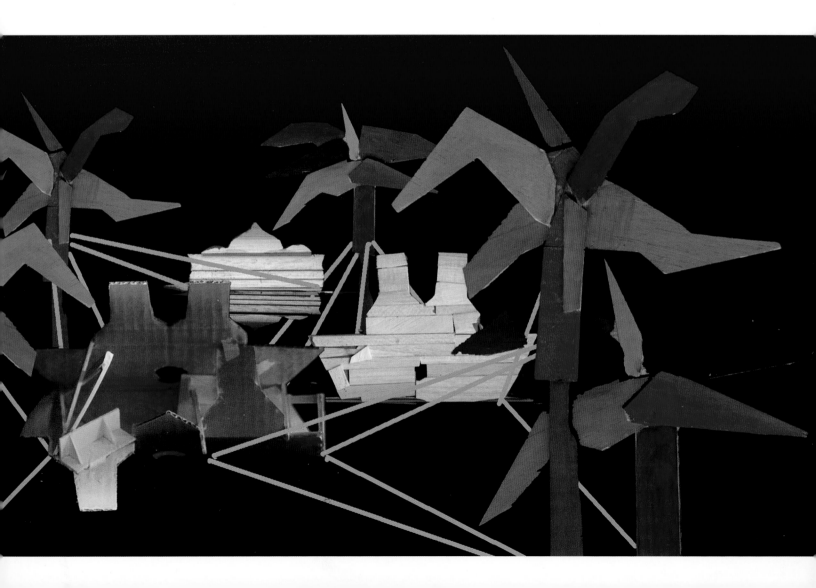

At the beach level there is an area that serves as a cooking space as well as the platform for a spiral-stair entry to the treehouse. There is also an area for the toilet and shower; these three functions are sheltered by the bamboo sail-like structures that surround each tree. All areas can be reached by the public space. Bamboo is connected to the units with joints that enable it to rotate in different directions.

If the wind pulls on the trees, the bamboo floor periscopes. The whole structure can expand and shrink. Units have shutters that can be closed.

Units can form a closed room by closing sliding glass doors between them and lowering the hanging roof. There are ramped corridors for going around and on top of the units. Movable roofs for the living areas can be lifted. The large surrounding bamboo "baskets" are for closing the area from view and sun.

GREENBERG: This project is well thought out in its integration of a modern "enclosure" surrounded by nature and indigenous forms and materials. It is also unique in its openness and privacy.

Hagar Arad and Tal Modai

This single-family treehouse is a sleeping box surrounded by open terraces. The box is a "breathing" skeleton that can be protected from heavy rains by rolling plastic sheets. There are three floors, and the overlapping surface between them creates the sleeping cube.

Each floor rests on a couple of beams connected to a pair of trees, stabilized by a set of four cables connected to a metal sleeve encircling the tree. Small platforms are placed on the edges of the main beams and carry the staircases. A system of cables passed through bamboo "columns" running between the floors and creating the cube help keep the structure stable.

Rainwater is collected from the roof. Shower water is recycled into the composting toilet.

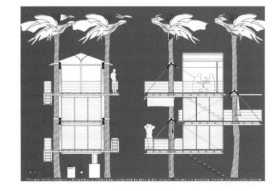

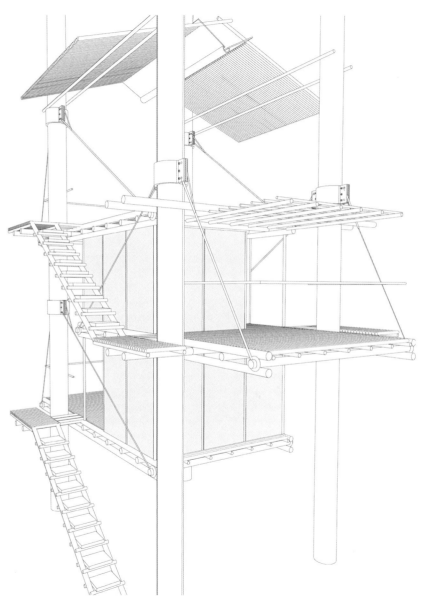

96 France

BAMBOO BEAM

This is a contemporary, low-technology alternative to the romantic image of the hut in the wilds. It avoids reference to palm trees and hammocks, but does not exclude natural materials and ecological choices. The chosen construction is the trussed beam made of bamboo, allowing maximum open space and minimum number of trees. Movable and variable partitions allow privacy and openness to the environment. The construction of this building, presented here with four planted palm trees, can also be easily adapted to a more natural configuration of trees.

DWELLING

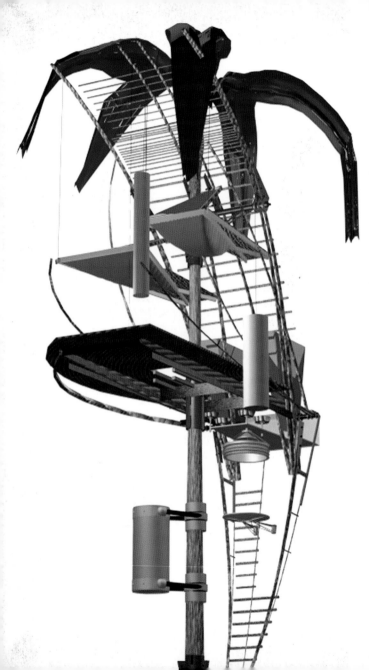

Marc Ackerson and Oisin Clancy
97 USA / Ireland
Top Ten Finalist, Second Place

A treehouse is neither a tree nor a house. It establishes a symbiotic relationship between the tree and the house. The intervention is interwoven within the tree. Its movement allows this relationship to fluctuate, blurring the edges.

Each building component has been developed to expand the concept of floating. The demountable, telescopic frame is easily transportable to any exotic location and is engineered to work in conjunction with the natural structure of a palm tree. This shelter's graceful form extends upward, uninterrupted by branches.

Like a bow and arrow, the treehouse balances tension and compression, constantly readjusting for a dynamic balance. Human and machine become one. The moment is tense. The space between the bow, the arms, the string, and the arrow is empty but very alive. The structure adjusts and flexes to accommodate variations in size, shape, and movement on the tree. The house and the tree form a union and begin to act as one, like a dancer. The treehouse's embrace of subtle movement seems to defy gravity.

From its tapered base, gesturing toward the ground, to its stargazing sleeping lofts suspended near the tree's canopy, the treehouse deftly combines the ergonomic comforts of a small dwelling with the ecological advantages of the tropics into a series of vertical zones. The stacked spaces include shower, toilet, kitchen, living space, and sleeping lofts as well as gas cooktop and hot water. Water tanks collect water for a gravity-based network while acting as counterweights for the structural movement and enclosure for the kitchen.

A limited and carefully engineered rotational movement around a pivot hinge at the base ensures the entire aerofoil-shaped form minimal deformation, torsion, and sway.

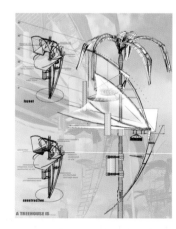

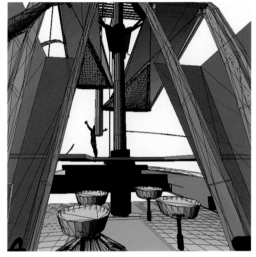

BURTON: ...kind of primitive, and the concept of kinetic movement is like poetry—quite beautiful.

GREENBERG: This is the one that becomes kinetic—the whole thing closes up to resist the typhoon.

FUNG: This looks like a trampoline diving board. Very insight-oriented.

GRINSTEIN: I love the way it closes its wings in the storm.

Kentaro Kimura, Toshio Ozaki, and Maki Yoshida

98 Japan

They first appear on the surface of the earth, gradually grow above our height, and eventually reach the sky, when they become strong enough for us to lean on. Multiple trees develop different relationships, such as pairs and triangles; spaces in treetops have infinite relations as the number of trees multiplies.

We decided to focus on the relationship between four selected trees, which stand in a square configuration. Natural rope was then placed around the four trees to create an enclosed space. The vertical space created by the rope was where we chose to place the floors. The ropes function as a filter from the surrounding environment.

Spending time in the treehouse can extend one's ability to notice spatial relations, and this ability can then continue to develop on one's return to the ground.

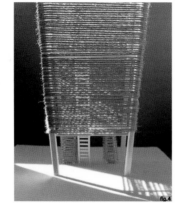

GREENBERG: The jury as a whole complimented and singled this project out for being unique in its adherence to a simple system.

GRINSTEIN: I like the transparency of it in the middle of the jungle, feeling enclosed but open. Not as serious as the others.

Acknowledgments

Acknowledgments and a whole lot of THANKS!!!!!!!!!!!!!!!!! go to:

Kirsten Kiser (Ki Ki) and www.arcspace.com for hosting the competition announcement; also web master Rimas Steponatis.

Bernad Perloff and Roger Webster, of Perloff/Webster for hosting production of this book and the competition, as well as for their advice, encouragement, consultation, and the use of their office/studio space and staff in Venice, California. Derek Plaza, for hours of computer magic, and office manager Nicole Siskind for her valued assistance. Special thanks to Roger Webster for his help, photography, and production assistance, not to mention philosophical dialogue during the course of the project.

The Jury: Elyse Grinstein, Ming Fung, Deborah Sussman, Pamela Burton, Stephen F. Jones, Billy Al Bengston, Ken Price, James Ganzer, Michael Garnier, Kirsten Kiser, and Master Chang.

Jim Ganzer for his original treehouse design that helped inspire the competition.

Joel Goldberg for faith and financial contribution. Richard d'Abo partner, friend, investor, advisor, art and life connoisoir. Abby Sher who generously donated gallery space for the competition at her wondrously creative Frank Gehry-designed Edgemar Center on Main Street in Santa Monica, California. Jim Ganzer, Sandy Sussman, and Flora for their help designing and hanging the show. Tao Ruspoli, for the documentary video coverage of the competition and for his computer. Thomai Hatsios for image management and Christine Sette for information management. Li Li, who got me to Hawaii and China. Jonah Greenberg, my son, who helped me build and encouraged me in designing the Big Beach in the Sky in China. Jayne Rockmill, Agent Extraordinaire.

At least half of the more than 100 winners whom I have met and have been so helpful, especially: Mark Ayers, Oisin Clancy, Marc Ackerson, Mireille Turin, Marcus Bach, Robert Garneau, Caroline Trolle, Marianne Bär.

(Buddha) Ling Hua Ping, who helped build the treehouses in China and has helped take care of them ever since. Steven Lao, Hainan office manager,

and Lu Huong. Rene Flack, who takes care of the treehouses in Hawaii when I am running around the world.

The County of Maui, and the State of Hawaii for encouraging me in my experimentation of treehouses. Adele Rugg, very special assistant to the Mayor of Maui, Alan Arakawa. Jim Dorian, the advance man from the Department of Business Economic Development and Tourism for the State of Hawaii, who helped me with my first contract with Nanshan Buddhist Culture Zone. Hawaii state senator, Kalani English. Zhang Hui, general manager and director of Nanshan Buddhist Culture Zone, as well as Hu Xaio Dong of BTC.

The County of Sanya and the Province of Hainan, PRC, for their support of the treehouses in China, particularly Governor Wang Xie Feng (now party secretary) and Mayor Chen Xu.

Zhou Ping of the Hainan Department of Foreign Affairs for being a friend in China and constant support of the treehouses there.

President Bill Clinton's White House staff and his Secretary of Commerce, Norman Mineta,

and Gregory Wong at the U.S. Embassy in Hanoi for their support of Treehouses in Vietnam, and Chairman Me of Hue Province.

Master Chang, who tells me I may never actually be a Buddhist or even totally Zen but have a good chance of being sort of a Taoist, as well as, of course, his sister, Than Lo. And Rosemary Shoong, my muse and more!

Andrea "Tarzan" Danese, my editor at Abrams, as well as Eric Himmel, Abrams' editor in chief, and designers Adam Michaels and Prem Krishnamurthy of Project Projects.

I would especially like to thank my mother, Edith Greenberg, and my uncle, Dan Greenberg, and Sid, who were such good friends and supporters of my dreams and who, from the beginning, believed in my crazy treehouse ideas.

Partners of future treehouse resort (village chief in center), Naviti Island, Fiji

To learn more about the designs and architects featured,
contact the author at www.treehousesofhawaii.com.

Editor: Andrea Danese
Designer: Project Projects
Production Manager: Maria Pia Gramaglia

Library of Congress Cataloging-in-Publication Data
Greenberg, David Lawrence, 1942-
    Treehouses in paradise : fantasy designs for 21st-century
    globe-trekking nomads / by David Greenberg.
       p. cm.
Includes bibliographical references and index.
ISBN 0-8109-5837-6 (hardcover : alk. paper)
    1. Tree houses—Hawaii—Hana Region—Design and construction.
    2. Tourist camps, hostels, etc.—Hawaii—Hana Region. I. Title.
TH4885.G74 2006
728'.5—dc22
                    2005027491

Printed and bound in China

10 9 8 7 6 5 4 3 2 1

HNA
harry n. abrams, inc.
a subsidiary of La Martinière Groupe

115 West 18th Street
New York, NY 10011
www.hnabooks.com

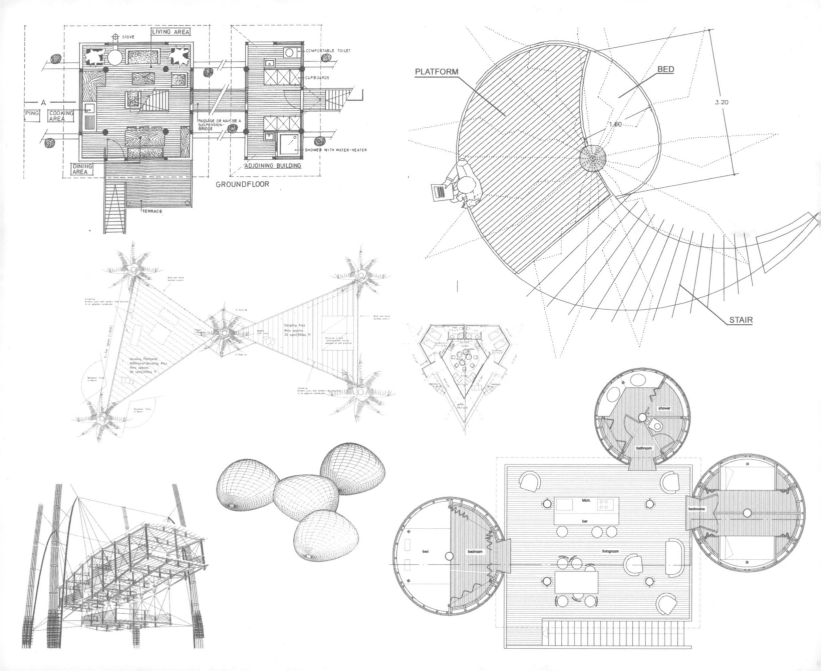

GROUNDFLOOR

STOVE
LIVING AREA
COMPOSTABLE TOILET
CUPBOARDS
CLEANING
COOKING AREA
DINING AREA
TERRACE
PASSAGE OR MAY BE A SUSPENSION-BRIDGE
SHOWER WITH WATER-HEATER
ADJOINING BUILDING

PLATFORM
BED
3.20
1.60
STAIR

kitch.
bar
livingroom
bed
bedroom
bedrooms
bedroom
shower
bathroom